BRACING FOR IMPACT

TRUE TALES OF AIR DISASTERS AND THE PEOPLE WHO SURVIVED THEM

Robin Suerig Holleran
and
Lindy Philip

Foreword by David Soucie

Skyhorse Publishing

Skyhorse Publishing books may be purchased in bulk at special discounts for sales promotion, corporate gifts, fund-raising, or educational purposes. Special editions can also be created to specifications. For details, contact the Special Sales Department, Skyhorse Publishing, 307 West 36th Street, 11th Floor, New York, NY 10018 or info@skyhorsepublishing.com.

Visit our website at www.skyhorsepublishing.com.

10 9 8 7 6 5 4 3 2 1

Library of Congress Cataloging-in-Publication Data is available on file.

Cover design by Rain Saukas
Cover photo by David Shenkin

Print ISBN: 978-1-63450-426-3
Ebook ISBN: 978-1-5107-0032-1

Printed in the United States of America

CONTENTS

FOREWORD

Yesterday, more than 100,000 people boarded airplanes, endured long security lines and uncomfortable flights, and then arrived at their destinations just as they had planned. The day before, and the day before that, hundreds of thousands of people purchased tickets on airlines or boarded private airplanes. Once onboard they read magazines, watched movies, or simply rested until their planes landed safely on runways. Day after day, thousands of successful flights add credence to the extraordinary safety of flight transportation.

Millions of people routinely accept the risk of death from what seem to be simple activities without giving them a second thought. Walking across the street, driving a car on the highway, and even riding a bicycle have exponentially higher risks than flying in an airplane; yet, flying often conjures up the fear of a horrific death in the public imagination.

In order to overcome the fear of flying, you have to fully understand just how safe it is to fly. The probability of dying in a tragic plane accident is so remote it can be difficult to comprehend. I call it the "11 million to 1" rule. The odds of being killed in an airplane accident are less than 1 in 11 million. With those favorable odds even the most skittish of flyers are willing to accept the risk.

However, when an airplane's engines do flame out and the plane plummets to earth from 30,000 feet, you are told to brace for impact. The 1 in 11 million statistic becomes real—you are the unlucky "1"!

Bracing for Impact tells the unimaginable personal stories of these unlucky ones—those who have suffered the horror of a deadly plane crash but who were miraculously given a second chance. The stories are captivating and enlightening, providing a perspective only survivors of airplane accidents can share, as well as a perspective that, as an accident investigator, I rarely have the opportunity to hear.

David Soucie
CNN Safety Analyst
Author of *Safer Skies*, *Malaysia Airlines Flight 370*,
and *Why Planes Crash*

INTRODUCTION: FLYING INTO FATE

For those who belong to the unenviable club of people who have survived a plane crash, talking about the experience is not always easy. Some, understandably, avoid the topic entirely because it can trigger an eruption of symptoms most commonly associated with post-traumatic stress disorder. Others can discuss the events of the crash with little effect, as though it happened to someone else.

Bracing for Impact is based on a series of interviews with survivors of both small plane and large commercial airline crashes. The book, more than a factual account of the tragedies themselves, explores the impact on victims' lives afterward. Despite the circumstances surrounding a crash, very few people walk away from such a life-changing event unscathed in some fashion.

Our intent is not to frighten anyone away from flying. Traveling by plane has evolved from being a special occasion, requiring passengers to dress accordingly, to an everyday part of many people's lives. There are very few of us who don't hear the faraway drone of a plane overhead at some point during the day.

The ambient buzz from plane engines has become such an integral part of our daily background noise that few even notice it. On the contrary, when the 9/11 attack on New York City and the ash clouds from Iceland's volcanic eruptions closed airports in the United States and Europe, respectively, it was the eerie, unexpected silence overhead that caught people's attention.

There's also a story behind the personal accounts documented here.

In 2007, as the popularity of Facebook exploded on the Internet, coauthor Lindy Philip was one of millions who jumped at the opportunity to connect with family and friends. But she was also intrigued by how Facebook allowed groups of like-minded strangers to come together.

Having lived through a plane crash herself, Lindy knew there were other survivors in this large virtual world. So, she created a private Facebook group and sat back to see what would transpire.

Slowly, other plane crash survivors found the online group and a loose network began to form. The Facebook group continues to grow, and while most members are not active on a regular basis, those with an unusual shared past support each other by expressing experiences, views, fears, and achievements within a community that shares one commonality.

As a freelance writer and plane crash survivor, Robin Suerig Holleran instinctively knew that there were other stories of survival worth telling. Having recounted her own story numerous times over the years, she realized firsthand that most people are curious to know how one recovers physically and emotionally from such an event.

Going down in a plane and knowing beyond a doubt that your life is about to end—and then surviving—is both exhilarating and traumatic. The terror is difficult to describe, as is the residual emotion that lives inside nearly all survivors for the rest of their lives. It's hard for anyone who has not been in a similar situation to fully appreciate the effect.

Robin's initial account of her harrowing experience received Honorable Mention in the *Writer's Digest* 75th Annual Writing

Competition. The contest attracted nearly ten thousand submissions, so to stand out at all was significant. But, during the time she wrote the two thousand-word article, Robin was also high on painkillers and strapped in a body brace from neck to hip—and cringes when she reads the essay today.

Initially, when Robin first had the idea for this book in 2011, she reached out to members of Lindy's Facebook group and asked if anyone was interested in being interviewed and sharing their experiences. Some agreed and others declined. Sharing was a very personal, and often brave, decision for those who participated.

As preliminary interviews gave Robin a feel for how her writing project might take shape, what struck her was how differently people coped with their tragedies. Robin was ready to send a book proposal to prospective publishers; however, she had a major health event that put the project on hold.

Fast forward to 2014.

Lindy knew that Robin was a writer who was toying with the idea of creating a book. Robin hadn't really thought about it much for a few years, but Lindy was insistent. She said the time was right and her intuitions were spot on.

The two formed a unique collaboration. Robin was the writer and Lindy was the force working behind the scenes to keep the Facebook group active, reaching out to members who were interested in being a part of the project, doing interviews over the next several months, providing editorial support and constructive criticism, being a motivator . . . and all the other "stuff" that needs to happen to make a project like this a reality.

Ironically, the two women had never met in person until about six months into the project because Lindy lives on the west coast of Canada and Robin resides in the New York City area. At last, Lindy came to New York with a couple of friends for a girls' weekend.

Robin hopped on a train and met Lindy for a drink in Penn Station. When they finally came face-to-face, they were wearing nearly identical outfits: black pants cropped at the calf and bright shirts splashed with shades of vibrant yellow, green, and red. Of course, they both laughed but the coincidence was just one of many they discovered along the way.

Lindy and Robin are grateful that Facebook brought them together and thankful to the members of the group and others who agreed to revisit the painful and traumatic events of their own crash incidents. Their stories of survival during and after the crashes are amazing and there were times when Robin had to put the writing aside because the stories caused her to choke up and break out in a cold sweat.

Through their unwavering resilience, all the survivors documented in this book have accomplished much despite often overwhelming odds. These individuals are living proof that, although it sometimes takes years on a bumpy, circuitous route, it is indeed possible to overcome significant physical and emotional setbacks.

This book, based on heartfelt interviews conducted in 2014, detailed research, and thoughtful fact checking, is the next chapter for those who participated. We apologize in advance for any inadvertent errors or omissions; each accident is based on the perspective of the person being profiled. It is certainly possible, and even probable in some cases, for two people at the same crash site to have had two entirely different experiences. In fact, three members of Lindy's Facebook group were all on the plane that crashed in Little Rock, Arkansas, one of whom is profiled in "Amazing Grace" (page 119).

People do survive plane crashes, but some people don't. In the time that Lindy and Robin started this book until the time that it was submitted to Skyhorse Publishing, major commercial disasters took the lives of nearly 300 in Ukraine, another 350 people in Southeast Asia in three separate accidents, 150 in Europe, and 116 in western Africa, as well as many others in smaller commercial and private plane crashes.

Bracing for Impact is dedicated to plane crash survivors as well as those who are left behind after an aviation tragedy. "Aftermath" (page 177) includes interviews with family members of victims who were able to funnel their grief to make the skies safer and to provide much-needed support to others struggling to come to terms with sudden loss from a plane crash.

LIGHTNING CAN STRIKE TWICE

Until her crash in 2005, author Robin Suerig Holleran had always assumed that she was safe in the air because the odds were in her favor. Her mother's stepfather was killed, along with forty-nine others, when a Pan Am flight crashed in the jungles of northern Brazil on April 29, 1952. Dan Radasma was the vice president of the Borneo Sumatra Trading Company in the New York office. He was coming home from a business trip to Rio de Janeiro when he died.

What are the chances of a second crash happening within the same family? In the case of the Suerigs, apparently pretty good. Besides Dan's untimely death, Robin and her father were in near fatal crashes and her brother fell from about three hundred feet when he was riding in an ultralight that faltered on takeoff. Robin still hasn't decided whether she, her father, and her brother were extremely lucky to have survived—or if a public service announcement should be made any time one of them boards a plane.

In addition, during the writing of this book, Lindy was introduced to a Canadian man who has survived three crashes on regional commercial airlines. And then there is the well-publicized case of Justin Hatch, a young man who survived two plane crashes. The first, in 2003, killed his mother and two siblings; the second, in 2011, took the lives of his father and stepmother just ten days after he'd committed to play basketball for the University of Michigan. Although Justin was in a coma for two months with a brain injury, fractured ribs, and a punctured lung, he was still able to score a point as a freshman in the last seconds of an exhibition game against Wayne State University.

Don't ever believe that lightning can't strike in the same place twice. It actually can—and does.

SOMEONE NEEDS TO BUY ME A LOTTERY TICKET

December 27, 2005
Cessna Cardinal
Locust Grove, GA

Robin Suerig Holleran. Photo credit Diane Terry.

As told by coauthor Robin Suerig Holleran.

Whenever my own plane crash comes up in conversation, three things happen: there's usually a moment of stunned silence, then an exclamation of "Oh my God!" followed by the inevitable question: "Robin, did you think you were going to die?"

The honest answer is yes; there was no doubt in my mind. After all, when you're cruising in a single-prop plane at two thousand feet and your pilot, and brother-in-law, tells you that there's a "problem" moments before the engine freezes, death isn't such a far-fetched concept.

Two days after Christmas, I learned the true meaning of the word "terror." It went far beyond anything I'd experienced in the first forty-five years of my life. Electricity soared through my veins as though I'd stuck a wet finger in an electrical outlet. It honestly wouldn't have surprised me if every hair on my head was standing straight up in the air.

I'd driven to Atlanta with my three kids to visit my sister Kathy during the holiday break. Soon after we got there, I asked her husband, Al, to take us to see his plane at a small airport nearby.

It seemed like a good idea at the time to get the kids out of the house and burn off some pent-up energy after the fourteen-hour drive from New Jersey. It was one of those crisp winter days when the sun is strong and warm on your face, but you still need a thick sweater.

My kids and their college-age cousin, Sarah, helped pull the four-seat Cessna out of the hangar and onto the tarmac. The boys, Patrick and Connor, had goofy grins and peppered Al with questions. My daughter, Morgan, on the other hand, was quiet and seemed a little apprehensive. I've always loved to fly and hadn't been in a small plane since college, when I used to fly for fun with a couple of friends who had their pilot licenses. This was the kids' first time.

As she sat in the back seat, Morgan's eyes grew as wide as saucers and her eyebrows rose impossibly high as the plane took off. I laughed. She was, after all, a twelve-year-old girl. The boys were pumped.

Patrick, Morgan, and Connor went up for a twenty-minute spin, at one point flying low enough so I could take pictures. At the end of their ride, the children disembarked and it was my turn. Sarah chose not to accompany me, openly admitting that she wasn't a big fan of small planes, and hung out with her cousins on the ground instead.

We took off just as dusk's first mellow pinks and purples began to stain the sky.

My brother-in-law, Al, wanted to show me a nearby fly-in community with a private landing strip called Mallard's Landing because he was contemplating moving there. Unfortunately, we never quite made it.

After only a few minutes in the air, Al said, "Something's wrong. We're losing power."

"Don't say that. It's not funny," was all I could choke out, the palms of my hands instantly covered in sweat.

As if on cue, there was a dull clunk, and the single propeller on the nose of the plane—just a few feet in front of my shocked and disbelieving eyes—froze in place.

An eerie silence engulfed the plane. We'd gone from a deafening engine roar to only a surreal *whoooooshhh* of air rushing past the window. The plane glided for a few painful moments and then the nose started dipping down toward the earth below.

Fear flooded every pore in my body. I couldn't believe this was happening. We were going down, and there was no landing strip or runway anywhere in sight.

Whoooooshhh . . .

Animal-like groans eked from my gut as I pressed myself back hard into the seat. I remember not knowing what to do with my hands. I wanted to hold onto something, anything, but at the same time I was telling myself no talking, no screaming, no touching anything or doing anything to distract Al because he was our only chance of surviving. Later, Al would tell me that I kept moaning, "No, no, no . . ." although I have no recollection of it.

Al called a mayday into the control tower at Tara Airport where we'd taken off and asked them to send someone to sit with my three kids and their cousin on the tarmac.

Whoooooshhh . . .

In the small, cramped cabin, with the engine so close it seemed to be in our laps, the only way Al and I could communicate was through the headsets we were both wearing. I learned later that my youngest son Connor was listening to the handheld radio Al had given him. When he told Sarah that something was wrong with our plane, she flipped off the radio and said, "I'm sure it's nothing."

I'm grateful that they weren't listening to the crash as it unfolded.

It was rush hour and traffic clogged the road below us. Perhaps it was his military training, but somehow Al maintained his self-control and never panicked as he aimed the plane toward an excavated field below us.

It took about two or three minutes from the time the engine stopped until we hit the ground, which might sound like a short period of time but I can tell you that it felt like an eternity.

As the plane continued to lose momentum and started falling from the sky, the strangest thing happened. I'm not a religious person and can count on one hand how many times I've been in church other than for weddings, but as we started rocketing toward the red Georgia clay below, a peculiar calm swept over me. It was almost a resigned acceptance of the inevitable. I was going to die and there was absolutely nothing I could do about it.

My life didn't flash before my eyes and, oddly, I didn't think of my kids or anyone else. Instead, with an almost detached curiosity, I stared at the ground rushing up to us and thought, *Well, I guess I get to find out for myself whether there really is a God. Wonder if I'll see a sign—a flash of light, perhaps—when we hit.*

We barely skimmed over the power lines. I was sure the plane was going to get snagged and send us into a cartwheel.

We slammed into the ground and there was a sickening screech of twisting metal. The wheels sank into the muddy ground, the nose jerked down, and the plane jolted to a stop.

Pain coursed through my body. Every inch from my neck to my hips screamed. I've broken more than my fair share of bones over the years and knew I was in trouble. The pain was through the Richter scale.

The plane was tilted to my side. I pulled the seat belt latch and pushed the door open, tumbling into the red mud still damp from a recent rain. I learned later that the wing was dipping into a foundation that had been dug for construction of a shopping center. Al couldn't see it from the air. Had we landed a few feet to the right, we surely would have slammed into the foundation's dirt wall and been killed.

The smell of the moist earth was somehow comforting. I wiggled my toes and knew I wasn't paralyzed. I just lay there gasping, trying to tamp down the pain.

In the back of my brain, I heard Al shouting for someone to call 911.

Suddenly a round, country-boy face appeared out of nowhere, inches from my face. Dressed in worn denim overalls, the man asked earnestly in a thick Southern accent, "Ma'am, ma'am, are you okay?"

I couldn't speak and I'm sure I gave him one of those *duh* looks that teenagers reserve for their parents.

This man whipped off his trucker cap and placed it across his chest. He laid his other hand on the top of my head and began an emotional discourse: "May the Lord please deliver this woman from her pain."

I shut my eyes. Maybe I really was dead.

The man jerked back and I heard Al yell at him to get the hell away.

As the sky grew dark, voices swirled around me, asking if there was any gas leaking. Several emergency medical technicians gently pulled me out from under the plane's wing. They cut off my sweater and rolled me onto a stretcher.

Holding a penlight in his mouth, a man in his early- to mid-thirties with sandy brown hair was trying to start an IV but my normally prominent veins had started to collapse from shock. I muttered protests; my fear of needles somehow still got the better of me. But then my muddled mind started to wrap around the fact that I was actually alive!

"What did you say?" He looked puzzled, to say the least.

"Someone needs to buy me a lottery ticket. This is the luckiest day of my life," I whispered.

He stared at me. I'm sure he thought that I had also been head-injured.

Sirens wailed and the *whoop, whoop, whoop* of helicopter blades burped overhead.

"Why are there helicopters?" I asked.

"We need to get you to a hospital," he said.

"But I was just in a plane crash. Can't I go in an ambulance? *Please*," I said between gasps.

"It'll take too long, it's rush hour," he said. And that was the end of that conversation.

We waited and the *whoop, whoop, whoop* got closer.

"Is that it?" I asked.

"No, that's the news helicopter."

Geez. I thought that only happened in Los Angeles.

It was dark and the technician sitting next to me in the helicopter reminded me of an astronaut with his helmet and flight jumper. He kept leaning close to look in my eyes—probably to see if I was losing consciousness.

Seconds after landing at Grady Memorial Hospital in downtown Atlanta, I was whisked off to the emergency room. A half-dozen masked physicians surrounded me, all peering down and firing off questions about where it hurt. Orders were barked for X-rays as braces were snapped over my torso and neck. The faces blurred and the pain thankfully receded with the first shot of morphine.

And then everyone was gone.

Just me and my painkillers . . . *wooeeee* . . .

On the other side of a flimsy blue plastic curtain, medical staff wrestled with a deranged patient as I stared at an arc of dried blood on the ceiling above me, wondering vaguely how it had gotten there.

Rocking in and out of consciousness, I heard conversation drifting past. "Another gunshot wound . . . plane crash . . . fractured back . . . Did he have a seizure and then hit his head or did he hit his head and then have seizure? Anyone know?"

A tsunami of nausea from the painkillers washed over me.

"Help, I'm going to get sick. Please, help," I squeaked.

No one came, so I tipped my head to the side and threw up all over the floor.

A woman (or at least I think it was a woman) peeked from behind the curtain, wrinkled her nose in disgust, and backed up the way she came. Occasionally, my head wobbled to the side of the gurney to check and, yep, my vomit was still on the floor—even hours later as I was wheeled to a different intensive care room.

My family showed up after making their way through the metal detectors, just as I was being wheeled out for an MRI. My daughter Morgan huffed when I kept asking if the plane was okay in my drug-induced haze. Like who cares?

The hallway was filled with people on gurneys. Some were handcuffed with an officer nearby, others were just waiting. I guess they sent me to Grady Memorial because it was a busy trauma center, but it was busy with mostly inner-city emergencies.

I lay quietly in the clunking imaging machine, my eyes glued to the blood splatter and flecks of gore on the inside of the tube.

Even to a layperson, the MRI images that came back told a frightening story. We'd hit the ground so hard that a vertebrae in my mid-back burst, sending bone chips out the sides to within millimeters of my spinal cord. The vertebra was now lopsided and, where the bone had cracked, it looked like a black magic-marker line had been drawn right through the middle.

The seat belt had cracked my sternum so I couldn't sleep in a bed for weeks because it hurt too much to breathe while flat on my back. I also learned about the pain of real-life whiplash, something I'd previously dismissed as a lawyer's disease.

As I drifted in and out of a fog during my stay at the hospital, my father, an emergency room doctor from Connecticut, sat in a chair next to my bed. He waited—and waited—to speak to someone on the elusive trauma team assigned to my case. But no one came. Instructions mysteriously appeared for the attending nurses, but not once during my four-day stay did a physician come in to talk to us or to examine me despite my dad's explicit requests.

This place was far different from the hospital where my father worked and that I'd come to know growing up. Not that I cared much at the moment (the drugs made sure of that), but the steam building behind my father's ears was becoming more and more noticeable, even to me.

At some point one night, the automatic blood pressure monitor that inflated and deflated on my arm slipped over an intravenous feed while I dozed. Awakening, I ripped off the pumping sleeve and pulled the bloody glob out of my arm. When a nurse finally did arrive, she gave me a snotty look, took the mangled mess, and, without cleaning the blood off my arm, reattached the blood pressure monitor. Even in my haze, this didn't seem right.

On the third day, my father asked me to sit. Then he asked me to stand. As I clutched his arm and teetered to my feet, still bound firmly in neck and back braces, a nurse rushed in.

"You can't do that! You need permission from the doctor," she said, obviously flustered.

"You go get permission if you want; I'm going to see if she can stand," my father growled.

Moments later, three people clustered at the doorway, looks of alarm spreading on their faces.

"She's leaving tomorrow," he said calmly, but without room for negotiation.

Equipped with a prescription for painkillers and an appointment with a neurosurgeon in Connecticut, we left the hospital on a quiet New Year's Eve morning. Although there had been a couple of nurses who truly seemed to care, the hospital stay had been a nightmare on top of the plane crash. All four intensive care rooms I'd been in had blood splatter on either the walls or ceiling. Encrusted remnants of food stuck on table trays and dirt and grime were smeared everywhere.

Several pilot friends of Al's offered to fly us home, but low-lying cloud cover filled with ice hung along the East Coast, eliminating that option for any plane without deicing capabilities. And the seats on commercial flights couldn't tilt back enough for me to make the trip in my Frankenstein suit.

Early the next morning, my father and I began the fifteen-hour drive home in a large rented SUV with wide reclining seats. Severe weather forecasts both in Atlanta and the New York area gave us a window of just one day to get home. Percocet was my very good friend during the ride.

Up north, we accomplished more with the neurosurgeon and radiologist in ninety minutes than we had over four days in the Atlanta hospital. Even better, I made it home to be with my children on my forty-fifth birthday.

My then-husband picked up our kids from school the day I got home without mentioning a thing. At the risk of sounding melodramatic, I have to say that the smiles and tears on their faces when they walked in the door was the best birthday present I've ever had. It was as though I was seeing them for the first time. If things had turned out differently, my only regret would have been not watching them grow up.

After coming so close to losing everything, I couldn't help but wonder why I'd been so lucky. A week before my accident, our little town had been rocked to the core when two cheerleaders—one of whom lived only a few doors away—were killed in a car accident coming home from practice on an icy winter night. About a month afterward, a young, recently married teacher I knew well was killed in another car accident.

I was consumed with guilt and kept feeling like I'd dodged a bullet that hit someone else. In the end, I decided that I must have been spared because I had unfinished business that needed to be taken care of. I needed to be here for my children.

As terrifying and difficult as the accident was, I'd do it all over again if it meant making sure that none of them would be on the plane when we went down—a possibility I can't even consider.

Al, on the other hand, said it never dawned on him that we might die.

"Part of pilot training is always preparing for the worst-case scenario. I wasn't that concerned until the plane stopped the way it did. We essentially went from fifty to zero miles per hour in seconds," Al said. "I admit, though, that it did take about six hours for my adrenaline to stop pumping and my heart rate to go down."

That's not to say that Al is immune to fear. While serving in the first Gulf War in the early 1990s as a firing battery commander, charging in front of tanks through fire and exploding artillery, there had been times when he'd expected to die. But he wasn't rattled on that cold, winter day outside Atlanta, at least not until he went back to the crash site the next day and saw the demolished plane.

It's a miracle that Al kept his cool and we survived. It's also a miracle that the engine didn't fail when my kids were flying in the same plane just minutes before me, or that the bone fragments didn't puncture my spinal column, or that any number of other things I don't like to dwell on didn't happen. Even Sarah turning off the radio at Al's first words of an emergency saved my children from the trauma of having to listen to their terrified mother's voice as she faced what she thought was imminent death.

Coming as close as I did to losing everything helped shift my priorities. I was given a second chance that could have just as easily not happened.

My ex-husband did the best he could do, taking on household and child-rearing tasks he was unaccustomed to. Tom is a good man, but we were not well suited for each other and had been drifting further and further apart.

As time passed, his anger about the crash, which I still don't understand, started percolating to the surface.

A few months after the braces came off, I needed to have a discectomy on my lower back and couldn't lift anything heavier than a gallon of milk for months, again needing help to complete even the simplest chores.

It took a little over two years after the plane crash for me to physically recover and for us to finally call it quits, although our marriage had been battered for quite some time. We are both very stubborn and neither of us wanted to be the first to admit defeat. But we also weren't doing our kids any favors by toughing it out. We were setting a poor example of what a married relationship should be. Although we never fought, there was zero affection—and very little communication—between us. At the dinner table, we spoke to the kids but rarely to each other—and then retreated to different rooms in the same house.

I didn't want my children to think this was normal.

When my divorce attorney asked what went wrong, I told her that the plane crash was a turning point in my life and in our relationship. I'd come dangerously close to learning just how fragile life can be and I didn't want to live the rest of it unhappily married—or making someone else feel the same way.

She told me that after the 9/11 attack on the World Trade Center, experts had predicted a strengthening of family units as people came together after a near-death experience. But that didn't happen. Many came home that day, looked at their husbands or wives, and finally admitted to themselves just how unhappy they were and asked for a divorce. They had come close to death and wanted something more from life. I did too.

About five years after the crash, I had an opportunity to climb Mount Kilimanjaro to raise money for a foundation founded when my stepbrother died of AIDS in 1991. When the idea

Tribune

Your local hometown newspaper

SUBSCRIBE TODAY CALL (908) 766-6960

Classified ads, call (800) 624-3684.

We want to hear from you To discuss news or sports: call (908) 879-4100

ROBIN HOLLERAN
28 DEAN RD
MENDHAM NJ 07945-1621

INSIDE

FRONT SECTION

Opinion 6-7
Obituaries 11-12
Sports 21-22

COMMUNITY LIVING

Happenings 13
Weddings 16

To advertise, call (908) 766-3900; to subscribe call: (908) 766-6960. One-year subscription in Morris County is $32. Newstand price is 75 cents. Out-of-area subscription rates available upon request.

Publication USPS 401980

(Please see Mendham on page 9)

Mendham woman survives Georgia plan

By PHIL GARBER
Managing Editor

MENDHAM - It was a perfect day for a leisurely plane ride over the Georgia countryside until Robin Holleran heard a pop and looked out to see the single propeller in the private plane had stopped.

The plane was cruising at about 1,500 feet, as the brilliant sun was beginning to set, when Holleran, a Dean Road resident, saw the treetops and the ground quickly coming closer. A terrifying few moments later, the pilot crash landed the four-seat Cessna Cardinal in a vacant area previously cleared to build new homes.

Holleran suffered a spinal compression fracture and assorted bruises after the Friday, Dec. 27, accident. She was taken by a Georgia State Police helicopter to Grady Medical Center in Atlanta and was discharged on Saturday, Jan. 31. She is now recuperating at the Greenwich, Conn., home of her father, Karl Suerig.

The pilot, Holleran's brother-in-law, Al VanLengen, suffered numerous bruises but refused medical attention.

Inspectors from the Federal Aviation Administration (FAA) are studying the wreckage to determine a cause for the crash. Holleran said the plane had been serviced just a few weeks earlier.

Fortunate Family

Holleran said in an interview on Monday that she considers herself lucky to have survived.

Fallen Craft

Robin Holleran, a Dean Road, Mendham, resident, was a passenger when a private plane crashed in the suburbs of Atlanta, Ga., on Tuesday, Dec. 27. She suffered a spinal fracture in the accident. The cause is under investigation by federal authorities.

Photograph by Morgan Holleran. Reprinted with permission of the Observer-Tribune/New Jersey Hills Media Group.

first surfaced, I wasn't even sure where Kilimanjaro was—but I went to Africa and climbed the twenty thousand feet to the top. It was hard. All of us were clobbered to some degree or another by altitude sickness, forcing three people in our group to turn back and an experienced porter needed to be whisked down on a stretcher after falling unconscious. My back will never be the same after the plane crash but I still made it to the top of Kilimanjaro despite my injuries.

And, in some warped way, I'm not sure I would have gone had I not been in the crash. In fact, I almost didn't go until I asked my then-thirteen-year-old son in casual conversation what he thought about his mom climbing one of the world's tallest peaks. His simple answer was, "Why wouldn't you?"

And he was right.

Al, too, made significant changes in his life soon after the crash, although he claims they were coincidental and totally unrelated. I'm not so sure.

"I essentially gave up a six-figure job for a six-dollar per-hour job. I'd retired from the Army after twenty-five years and had spent the last ten years as a firefighter with Booz Allen racing anywhere from Detroit to Abu Dhabi to put projects back on track," Al says. "But, I'd always loved to fly. Even as a kid, I would tie boards on my arms and jump off the roof."

Al never flew in the military. When he joined, the Vietnam War was winding down and many pilots were on the market. He got a private pilot license in the mid-1970s while stationed in Germany, but the arrival of his first child and the responsibilities of raising a family made it impossible to keep it up until much later in life.

So, he left high-powered consulting to become a full-time flight instructor. At the time of this writing, Al is a safety inspector for the Federal Aviation Administration (FAA).

Al recorded in the narrative section of his NTSB Pilot/ Operator Aircraft Accident Report:

> . . . there were no indications of any problems during the run-up procedures to the climb from the runway to 2000 ft. MSL. I was staying low and slow so she could see the area, 2000 MSL 120 knots. We were about three miles from Mallards when I notice [sic] the engine did not seem to be producing power. I was at 22 MP and 2300 RPM when I notice I had to keep pitching up to hold altitude. I checked all instruments and they were in the green, oil pressure, oil temp, Head Temp, all were normal. So I said we were turning back and headed towards 4A7. I increase RPM's and MP and was getting about 90 kts with a 100 FPM climb. Within one to two minutes after I initiated the turn back towards 4A7, OP went to zero and the engine quit, the propeller seized and stopped. I turned toward an open area and was looking at Mallards but knew I could not make that. I immediately declared an emergency on the 4A7 common frequency and asked someone to gather up the three children who were waiting with my daughter-in-law [Sarah]. I saw the best open area to my left, a newly excavated area that appeared relatively flat and smooth. I banked into it, set up for touch down, as the main wheels touched, the nose came down and the aircraft came to an abrupt stop. My sister-in-law exited the Aircraft out the right door, and said her back was hurting; I made her stop and stay on the ground under the right wing. I also exited out the right door. A Henry County police officer observed the landing and was on the scene immediately. He assisted in directing emergency personnel to the site. EMS airlifted my sister-in-law [Robin] to Grady Hospital.
>
> The entire landing was less than 60 feet. Damage to the aircraft is major by looking at it. No aircraft parts separated from the aircraft. This area is in the North West Corner of Georgia

Highway 155, and Hampton/Locust Grove Road Intersection. The field was a large open area but extremely wet, soft and muddy. There were some holes and depressions throughout the field. There were power lines and trees surrounding the field on the east and south, the direction I came from.

There was plenty of fuel in the tanks. There were no signs of oil leaks or oil around the cowl. There was plenty of oil in the engine, at 4 qtrs. by the stick. I am unsure of the cause of this at this time. They have taken the plane to Atlanta Air Salvage in Griffin, Ga. to further investigate.

After the accident, I went as a walk-in to Fayette/Piedmont Hospital in Fayetteville, Ga. I had minor facial cuts and bruising. Some lower back pain and minor aches and pains. I was released that evening. My sister-in-law was air lifted to Grady Hospital for diagnosis . . .

Photo credit Associated Press. A man is dwarfed among the twisted wreckage of a Convair 240 that crashed in a wooded area. The plane was carrying members of the Lynyrd Skynyrd band.

FREEBIRD

October 20, 1977
Convair 240
Gillsburg, MS

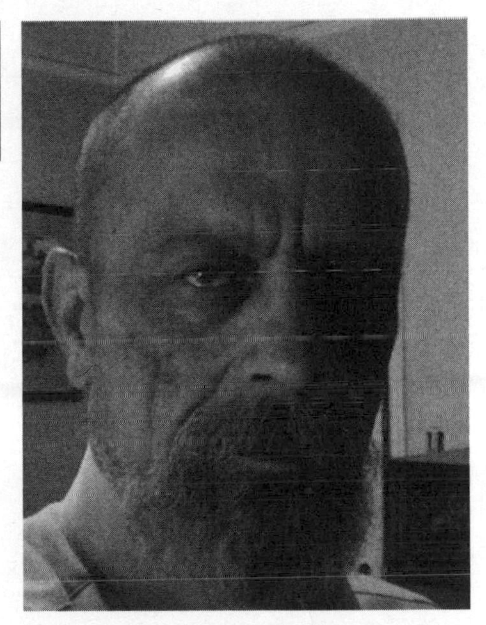

Marc Frank. Photo credit Marc Frank.

At twenty-four years old, Marc Frank was living the life. The legendary Southern rock band Lynyrd Skynyrd had just released a new album—*Street Survivors*—that was expected to soar to gold. Marc was heading out with them as a crew member of the band on their most ambitious tour to date.

Lynyrd Skynyrd (formed by five friends during their high school years and named after their nemesis, gym teacher Leonard Skinner, who harassed the students for their long hair) skyrocketed to fame in 1974 with the release of the still-famous song "Sweet Home Alabama." Lead singer Ronnie Van Zant referred to the band's style as "Southern raunchy roll" and was obviously proud of their rough-and-tumble reputation when he said, "The other bands are just as bad, but we go to jail more."

The 1977 tour was going well. But the transportation between concerts was a different story, especially during the five days leading up to October 20.

In a deep baritone rumble that one would expect from someone associated with Southern rock, Marc explains, "I've always been a white-knuckled flyer but everything was going fine in the beginning. The first pilot was an ace World War II pilot who knew the band well from previous tours."

Partway through the tour, the leasing company changed pilots and planes and according to Marc, there was a bad feeling about it from the beginning. Nothing you could necessarily put your finger on; it was subtler than that, but everyone felt it.

Problems started the moment the two new pilots arrived in Statesboro, Georgia, in a thirty-year-old Convair 240. Ronnie was sitting in the back of the plane across from Marc during the flight to St. Petersburg, Florida, and could feel wobbling in the tail as the pilots fiddled with the instruments.

Marc remembers Ronnie shaking his head and grumbling, "I really don't like this at all."

As the band was about to depart St. Petersburg, the pilots realized that they had forgotten their billfolds at the hotel and made everyone wait while they ran back to get them. Marc didn't think much of it at the time, but in hindsight, it was just one more hint that something was awry.

A couple of days later, mid-way through the flight to Greensville, South Carolina, Marc glanced out the window to see ten-foot streams of flames shooting out of the right engine like a sputtering blow torch. This sparked a lot of chatter in the plane as everyone asked each other if they saw what was going on. There was a torching problem in the right engine and it was becoming more obvious to Marc that the pilots didn't know the aircraft very well at all.

"Everyone on the plane wanted to turn around and go back," says Marc.

After Greensville, the band was scheduled to perform at Louisiana State University in Baton Rouge where ten thousand fans were expected. Marc was the first person to arrive at the airport, so he watched the pilot poke around inside the engine.

"I kept wondering why the pilot didn't just replace the damn part that was causing problems. When I asked him about what happened with the fire, he just shrugged it off," adds Marc.

Some of the passengers were still so unnerved after the incident with the engine fire the day before that they got together before the flight and discussed other options to get to Baton Rouge. One person made a reservation on a commercial airline but at the last minute joined the rest of the group on the troubled plane. Cassie Gaines, a singer in the band, talked with Marc about riding in the equipment truck instead of the plane, but Marc knew Cassie would never let her younger brother (the band's guitarist and song writer Steve Gaines) get onboard without her.

Despite the warnings, Marc doubted that anyone ever expected to crash. Still, he pointed out that any one of them could have said no during the several days leading up to the crash. But they didn't.

"Most people don't even remember what they did the day before they were in a plane crash, but I relive each moment of all five days leading up to the accident," says Marc. "It's hard."

The weather on that mid-October day, just five days into the tour, remains vivid in Marc's memory. It was cool, crisp, and clear—an all-too-perfect day for flying. Yet, anxiety was escalating within everyone. In an ironic twist of fate, a survivor would later claim that Ronnie Van Zant's last words before climbing onboard the plane were, "If it's your time to go, it's your time to go."

"Nothing was said about what had happened the day before, but the undercurrent of tension was palpable," he adds. "The mood on the plane sucked, and, in hindsight, the flight had a Buddy Holly-feel to it." Buddy Holly died in a plane crash in 1959, along with fellow musical stars Ritchie Valens, and J. P. Richardson (The Big Bopper) on what is often referred to as the "The Day the Music Died."

Despite the strain, there were no problems with takeoff and people started to relax and play poker. After the crash, rumors circulated that the pilots were at the back of the plane partying with everyone else, but Marc says that wasn't true. It was, in fact, a very low-key day for the band.

Then, somewhere over Mississippi, Marc looked out his window and saw fuel spraying in an arc out of the right engine. His initial thought was that the pilots were trying to transfer fuel from one engine to the other. He then grew concerned—he knew the plane might have the ability to fly on one engine, but not when it was as heavily loaded as it was then.

Suddenly the engine started sputtering and the plane began to jerk, almost like a car running out of gas, but at a harrowing nine thousand feet in the air.

The pilots had just called Houston Air Traffic Control Center to report that the plane was running low on fuel and to request radar vectors for the nearest airport. Marc says the plane heaved on its side and made a terrifying 180-degree turn with one wing turned heavenward and the other pointing straight to the ground below—in a desperate attempt to reach a small airport they had just flown past.

Things then spiraled out of control. Everyone ran around the plane in a panic, trying to strap themselves into their seats.

Marc recalls the strange sensation of being inside the plane one moment, hearing the drone of the engine, then suddenly being surrounded by deafening silence as the plane finally ran out of gas. All that could be heard was the soft whistle of air caressing the fuselage and the murmurs of people praying around him. When Marc stole a glance out the window, he could see treetops about five hundred feet below.

"There's no worse feeling. Plane crashes are as bad as you think they are," Marc says. "All I could think about were my parents and how my life was about to end. That this was it."

The plane scraped the treetops and then slammed into a heavily wooded area near Gillsburg, Mississippi, just eight miles from the airport the pilots were so desperately trying to reach. Seats broke loose and spewed around the inside of the plane.

People living nearby described the sound as something like a car skidding on gravel followed by the horrifying noise of wrinkling metal before the rumble of impact. With no fuel left in the tanks, an explosion was avoided when the plane splintered into pieces.

"I will always remember the sound of popping metal and the smell of the tree limbs that the plane dragged with it and that were now falling down around us on the ground," adds Marc.

There was silence for a moment, until the moaning started. Steve Lawler, another crew member who was sitting next to Marc, asked, "What happened?"

Marc replied, "We fucking crashed."

The scene was very surreal. Moments earlier, the plane had been flying in what seemed like broad daylight above the clouds.

But on the ground, in the early evening, everything around Marc was pitch-black. He describes the sensation as "moving between two different worlds."

Steve, crew member Kenny Peden, drummer Artimus Pyle, and Marc struggled out of the crumpled plane by crawling through a hole that had ripped open when the tail section separated from the rest of the fuselage. Somehow Artimus had managed to get his bearings by looking out the window as they were going down, so he knew the direction to a nearby farm. Air traffic controllers had lost contact and were unsure of the plane's location. The world-famous band had crashed in the middle of nowhere.

Marc was in shock.

"I felt like I was inside one of those plastic tumblers people drink out of. I could talk. I could see. But I didn't feel like I was a part of what was happening around me," he explains.

Kenny, too hurt to move, propped himself against a tree while Steve, Artimus, and Marc trudged through the dense underbrush of the Mississippi swamp for about forty-five minutes to the farm Artimus had spotted. Marc is sure that the sight of the three of them staggering out of the woods, wounded and covered in blood, forever changed that farmer's life.

The local man was working in the fields when he heard the crash in the woods. Spooked, he sent a warning round from a shotgun into the air, and the story was later exaggerated to say that the farmer had shot at the three survivors, but he hadn't.

"That farmer was from the country and had grown up with guns. If he'd wanted to shoot us, he would have," Marc says.

Artimus ran up to the farmer and yelled that they had to help the others get out of the plane, before lunging into the man's trailer to use the telephone.

Marc and Kenny were the first to be taken to the hospital, while Artimus went back to the crash site to help with the rescue efforts. It was while Marc was watching the news on the television in the hospital that it hit him hard. Six people he knew and cared about were dead.

News reports described helicopters bringing doctors to the scene and lighting the crash site with floodlights where some of the survivors were trapped inside screaming for help, "Get me out of here. Help, get me out."

It took more than three hours to pull the survivors and the dead out of the swampy terrain. Trucks got stuck in the mud and the responders had to wade through a waist-deep, twenty-foot creek with stretchers to reach the injured and dead. Marc says that there was controversy about the rescue efforts, but this had been a different time in the 1970s when rescue operations and technologies were not as advanced as they are today. Everyone had done the best they could.

"The real heroes that day were the Gillsburg Volunteer Fire Department," says Marc. "It was dark and there were no lights. They went in without hesitation, not knowing what to expect. It was a terrible scene, and I heard that some of the rescuers later struggled with post traumatic stress disorder."

Of the twenty-six people onboard, six were killed and twenty were injured, many severely. Among the dead were lead singer Ronnie Van Zant, guitarist Steve Gaines and his singer-sister Cassie Gaines, the two pilots Walter McCreary and William Gray, and the assistant road manager for the band, Dean Kilpatrick.

Rumors about the circumstances surrounding the crash started almost immediately. Autopsies of the pilots proved that there were no drugs or alcohol in their systems, despite accounts that they had been snorting cocaine the night before. Reports of drugs and money being found on the plane persist even today, even though the county sheriff at the time said that they had only found some loose bills scattered about, presumably from the poker game, and "just old drug store medicine."

Photo credit Associated Press. The wreckage of the Convair 240.

Tour manager Ron Eckerman claimed that $1,100 in cash used to pay tour expenses was missing from the briefcase that was later returned to him. Additionally, the lease between Lynyrd Skynyrd and the plane company contained a clause stating that the "lease shall hold lessor harmless in any event that drugs or narcotics *of any kind* should be brought aboard this aircraft for any purpose," which continued to fuel speculation even though it seemed obvious that the clause was intended to protect the leasing company from the transportation of illegal drugs.

Ronnie Van Zant's widow honored her late husband's wishes to have a closed casket so that people wouldn't be "gawking" at his body. This then led to stories of him being disfigured or decapitated by the head injury that took his life. According to his widow however, he had a bruise roughly the size of a quarter on his temple and looked like he was merely asleep when he was buried.

And still, twenty-three years after their death, Ronnie Van Zant's and Steve Gaines's bodies were vandalized in 2000 and subsequently moved to an undisclosed location. Although not confirmed, it is believed that the vandal was trying to confirm a long-standing rumor that Van Zant was buried wearing a black Neil Young T-shirt. The Lynyrd Skynyrd song "Sweet Home Alabama" with the lyrics, "Well, I hope Neil Young will remember / A Southern man don't need him around anyhow," is considered the band's rebuttal to Young's song, "Southern Man," bashing the South for being racist.

It would later be reported that Aerosmith had rejected the same plane earlier that year because they felt that the aircraft was not up to standards; they were also concerned when an Aerosmith representative caught the pilots swigging down a bottle of Jack Daniels during an inspection of the plane.

Three days before the tragedy, the *Street Survivors* album was released with cover art showing the band standing in flames, an ominous foreshadowing. MCA Records snatched the albums with the original cover back from the market and replaced it with new artwork out of respect to the band and their family members. *Street Survivors* would soon become Lynyrd Skynyrd's second platinum album.

Shattered by the death of their lead singer and the injuries suffered by the other survivors, the band fell apart. Some went on to form or join other bands, but none came close to the success of Lynyrd Skynyrd. Eventually, the band reunited and was joined by Ronnie Van Zant's younger brother Johnny as lead vocalist.

But tragedy seemed to follow the remaining band members. After the crash, guitarist Allen Collins's wife died after a miscarriage and Allen was said to have gone on a "death mission," overindulging in alcohol and drugs and driving anything with a motor way too fast. In 1986, he was paralyzed after a drunk-driving accident that also killed his girlfriend (he would eventually die in 1990). Two of the widows sued the reformed band for violating an agreement not to exploit the Lynyrd Skynyrd name for profit. Bassist Leon Wilkeson died in 2001 when his liver quit after a hard life of partying and keyboardist Billy Powell died in 2009 at the age of fifty-six from a suspected heart attack. Drummer Artemis Pyle damaged his leg so badly in a motorcycle accident that he was left with one shorter than the other. Craig Reed, a longtime member of the road crew, retired in 2005 and gave up drinking after catching hepatitis C. Reed collected the personal items from the plane that no one wanted and became the unofficial authenticator for Lynyrd Skynyrd memorabilia, which he sells on eBay to help pay his medical bills.

As expected, the cause of the crash was fuel exhaustion. The flight plan filed by the pilots before takeoff was also wrong, stating that there was five hours of fuel onboard even though investigators' calculations showed less than three hours of fuel left when they'd taken off.

According to the National Transportation Safety Board (NTSB) report, the pilots were running the right engine with the fuel mixture control in the auto-rich position to compensate for the problem with the right engine. But running the engine on the auto-rich setting caused it to sporadically torch and to burn an extra twenty-five gallons of fuel per hour that they didn't have.

The NTSB report states:

> The crew was either negligent or ignorant of the increased fuel consumption because they failed to monitor adequately the engine instruments for fuel flow and fuel quantity. Had they properly monitored their fuel supply and noted excessive fuel consumption early in the flight, they could have planned an alternative refueling stop rather than attempting to continue the flight with minimum fuel. In addition, the Safety Board believes the pilot was not prudent when he continued the flight with a known engine discrepancy and did not have it corrected before he left Greenville.

It took months for Marc to heal physically. His settlement from the charter company was a grand total of $28,000, and he jokes that he would have come out better if he'd fallen into the meat display at Walmart.

While some of his fellow passengers struggled with survivor's remorse, Marc says he didn't feel guilty for living through the crash; he just wished more had made it out safely. The loss of life was terrible and he believes that Lynyrd Skynyrd was poised to go places that no other American band had ever gone.

"They had what it took and everyone could feel that they were going to be great," he observes.

After Marc recovered, he started working for rhythm-and-blues bands, including The O'Jays, and country musicians. Southern rock had almost killed him and he never went back.

For years after the crash, Marc's dreams always involved the same two scenarios: Either Artimus or his mother flying a plane at tree level—or Marc in the back of a plane trying to reach the cockpit but the door always opening into an area resembling the space between the cars of a train. If he talks too much about the crash, even today, these dreams return to invade his sleep.

It took Marc two years before he could get back on a plane to fly from Detroit to New York City and then on to South Africa with a rhythm-and-blues band. He honestly didn't know if he was going to make it. The thought of the door closing, committing him to an eighteen-hour flight, was nerve wracking.

Over the years, Marc has bumped into a few other people who have been in plane crashes and he visited the Buddy Holly crash site after his own accident. Otherwise, he doesn't think much about it other than on anniversaries of the crash. And sometimes the memory of that day knocks him sideways without warning. It just happens.

Marc, who isn't a religious person, says that if he was saved for a reason, he hopes he figures it out soon because, at sixty-three years old, he "doesn't have a whole lot of time left."

He summarizes the ordeal as such: "It happened. I got on a plane, and it crashed. It's that simple. Life is a crapshoot. I survived because I was sitting in one place and others died because they were sitting somewhere else."

IT WAS A CRAZY TIME

The same night as the Lynyrd Skynyrd crash, a Frontier Airlines flight was hijacked in Nebraska by Thomas Michael Hannon, who boarded the plane with a shotgun and forced it to fly to Atlanta.

Hannon had been arrested for bank robbery and was out on bond. He demanded three million dollars and the release of his alleged accomplice and lover who was still in jail. Hannon eventually freed all of the hostages and shot himself in the chest.

In the 1970s, airport security was not as we know it today. Passengers simply walked up to the plane and showed their tickets to often-harried stewardesses as they walked onboard. Announcements were made just before takeoff asking people to make sure they were on the right flight. It was not unusual for passengers arriving at Dulles Airport in Virginia to think they were landing in Dallas, Texas.

DOUBLE JEOPARDY

Marc Frank grew up next door to the woman Artimus Pyle would eventually marry and he was one of the ushers in their wedding.

Six years before the fatal Lynyrd Skynyrd crash, Artimus's father died in a mid-air collision in Albuquerque, New Mexico. The two of them had been taking flying lessons, and his father had started flying solo.

Artimus was nearing the end of a four-year commitment with the military and had just signed-up for candidate's school to fly jets but lost interest after his father died. Because he was the sole surviving son, the military released him early, albeit only about a week before he would have completed his stint anyway.

So instead of becoming a Marine Corp officer piloting fighter jets, Artimus became a drummer in the Lynyrd Skynyrd band. It took him more than a year to recover from the 1977 plane crash but he continued playing music and, in 1980, narrowly survived a horrific motorcycle accident.

UP IN FLAMES

February 8, 1976
McDonnell Douglas DC-6/YC-112A
Mercer Airlines
Van Nuys, California

Guy DeMeo. Photo credit unknown.

As soon as he could, Guy DeMeo signed up for a local Air Explorers youth group, an offshoot of the Boy Scouts, and began taking lessons to learn how to fly. He'd always been fascinated with airplanes and his mother took him regularly to watch them take off and land at the local airport. According to Guy, Lockheed Corporation, the aerospace company now known as Lockheed Martin, was a local employer that helped sponsor the Air Explorers group by providing ground training, organizing volunteers to teach flying, and paying for a portion of the air time.

Photo credit Boris Yaro. Firemen scatter after saw ignites gas fumes at crash site of DC-6 in Van Nuys. Three trapped crew members of Mercer Enterprises DC-6 charter plane died. Ten firemen were injured.

When he was sixteen years old, Guy considered himself lucky to get a job cleaning hangars and doing odd jobs for a small, now-defunct charter company called Mercer Airlines that ran junkets from southern California to Las Vegas for people wanting to spend a day gambling in Sin City. Guy's claim to fame at the time was that he handled luggage for the band Pink Floyd.

On this particular February day in 1976, the plan was for Guy to fly in the plane for fun—a quick fifteen-minute trip from the Hollywood-Burbank airport (now known as the Bob Hope Airport) to Ontario, California, where the plane was scheduled to pick up some passengers. That same day, Guy was going to drive back to Burbank with a friend who was the ticket collector at the Mercer Airlines gate.

It had been raining for most of the week and there was some discussion about cancelling the trip because the weather wasn't letting up. It was still raining heavily when they took off.

"My dream at the time was to become a pilot and I was fortunate—or so I thought I was—to be able to go for rides in planes because I worked for the company," says Guy. "We were flying a DC-6 that day but I'd also ridden in their DC-3s and there had never been a problem."

Even in the 1970s, Mercer Airlines flew old planes. The DC-6 that Guy was riding in was designed near the end of World War II as a military transport plane and later modified for use in the commercial industry. The YC-112A being used this day was a DC-6 prototype that, according to Guy, was actually the first of its type built by Douglas Aircraft Company and it only had a couple hundred more hours before it was to be retired to an aviation museum.

It didn't quite make it.

The four-engine plane could carry about eighty passengers. Sometimes, the pilots allowed Guy to sit in the jump seat in the cockpit but, for whatever reason, this time they asked him to sit in the cabin with the two flight attendants. Besides the three pilots, Guy, and the other two Mercer Airlines employees, the plane was completely empty.

They took off and were only about one hundred feet in the air when there was a loud explosion and the plane started shaking violently. Guy thought it was one of the engines backfiring, which happened with some of the older planes, but was surprised at how fierce the backfire had been.

The flight attendants looked at Guy and asked incredulously, "What was that?"

Guy was sitting on the left side of the plane and looked out his window. The two engines on his side looked fine. He got out of his seat and walked to the window on the other side of the cabin. Horrified, Guy could see that the engine closest to the fuselage was missing. It had literally fallen off the plane. Witnesses on the ground reported seeing a flash of fire, dark smoke, and pieces of the plane falling from the sky.

"I thought I was going to die," says Guy. "To keep myself calm, I kept telling myself that the plane was built to be able to fly on two engines, but I was getting claustrophobic. All I wanted to do was jump out. To get out."

There was no way for him to know just how prophetic the thoughts racing through his mind would turn out to be.

He rushed back and buckled himself into his seat. He told the young attendants, both in their mid-twenties, that the engine had fallen off. They just stared at him, speechless and frozen in disbelief.

The pilots called in a mayday and the Hollywood-Burbank controllers told them that they could attempt a landing but might collide with the debris from the lost engine that littered the intersection of two runways. The pilots thought they could maneuver around the obstacles and land safely. Guy felt the plane touch down and heard the familiar revving sound of what he thought were the engines reversing to slow down. He let out a huge sigh of relief. All was good; they were on the ground in one piece.

Abruptly, Guy felt the force of the plane accelerating again. He thought it might be his imagination, but he was wrong. To his dismay, Guy realized they were going back up again.

In the recordings analyzed by investigators, the first officer was heard telling the captain that the propellers were not going in reverse—in fact, they were accelerating. They tried to use the brakes, and when that didn't work, the first officer shouted, "Get your air, get your air."

Maximum power was applied and the plane only cleared a blast fence and a large boulevard at the end of the runway by thirty feet.

When the plane turned right, Guy knew they were heading to Van Nuys Airport about ten miles to the west, where he was taking his flying lessons. What he didn't know was that Van Nuys Airport had a longer runway and the pilots were hoping that they would have a better chance of rolling to a stop there, since apparently there was no way of stopping the plane with the engines or brakes.

In a bizarre twist of fate, about half way to Van Nuys, an engine on the other side of the plane died and that propeller stopped turning. The four-engine plane was now down to just two engines, exactly the imaginary, never-could-happen, worst-case scenario Guy had used to calm himself down when the first engine fell off. The plane was losing power and struggling to stay aloft.

Unbeknownst to Guy or anyone else on the plane, a propeller blade had come loose, causing an imbalance that instantly tore the engine from the plane. The blade then whipped across the fuselage, severing the hydraulic line that controlled the nose gear, the pneumatic lines for the emergency air brakes, and much of the electrical wiring that provided propeller control, before ripping into the second engine that had just stopped running.

They were in serious trouble.

"I was told we were only a few hundred feet above ground, and some witnesses said it looked like we were skimming the tops of trees," says Guy. "When I looked out the windows, there were buildings that seemed as tall as we were high."

In the cockpit, a recording captured the pilot saying that the windshield wipers weren't working, clearly adding to the problems he was having navigating the plane in the rainy weather. The flight engineer explained that this was happening because the plane's hydraulic system was no longer working.

As the plane approached Van Nuys Airport, Guy was stunned to see how the crippled DC-6 continued to slide downward. They barely cleared the I-405 Freeway. He said his prayers and stared out the window, trying to grasp what was going on. There was no denying the situation; Guy crouched over in the crash position with his head in his lap and his hands on top.

The pilots knew they weren't going to make the airport and focused on Woodley Municipal Golf Course about one mile south that was fortunately closed due to the rain. The plane pummeled the ground, bounced into the air, came down again with a thundering crash, and rebounded back up into the air again. The plane then smashed into the twenty-four-inch concrete foundation of a building under construction that the pilots probably couldn't see without the windshield wipers working.

"There was this God-awful noise, the crunching of metal that seemed to last forever. It's so hard to describe," adds Guy.

When it finally came to a stop, Guy's first thought was that he was dead. He lifted his head and could see that the roof of the plane had collapsed and was folded like an accordion on the floor of the cabin just three rows in front of him. Seats had torn loose and were scattered everywhere. The overhead bins had detached and fallen on the flight attendants still strapped in their seats, but who were relatively unscathed. Guy says it was a miracle that the plane didn't burst into flames when they'd smacked into the ground.

He yanked open a nearby emergency door. The flight attendants were pinned in their seats and struggled to get out. The slide had fallen from the fittings on the door during the crash, so the three of them jumped to the ground which was now only about five feet below because the plane had collapsed onto itself. Fuel was spilling everywhere. The smell was overpowering—and terrifying.

The flight attendants were shaken but somehow convinced Guy to try to reach the pilots. In a stupor from shock, Guy climbed back into the plane but there was no way to reach the cockpit; the front of the plane was completely blocked by the wreckage. When the front landing gear hit the concrete pad, the force of the impact curled the nose of the plane under the cockpit, killing all three pilots instantly, Guy would later learn.

Guy still shakes his head when he recalls that the first person on the scene was a news reporter with a camera. Previously alerted of the emergency, a California Highway Patrol helicopter and the Los Angeles Fire Department were also there in seconds.

The fuel lines had ruptured and were spilling airplane fuel in the front of the wreck. No one knew at this point if any of the pilots were alive or dead. To get into the mangled cockpit, special fire-retardant foam was sprayed around the nose and wings before one of the fireman started using a rotary saw to separate the cockpit from the fuselage. Rotary saws had been tested without problems for emergencies like this, but the blade suddenly sent a shower of sparks when it hit a steel bolt in the aluminum body.

Whoosh.

The aircraft and about two dozen firemen were engulfed in fire. Some of their clothes ignited as they scrambled to get away from the blazing plane. Other firefighters grabbed the ones who had fallen and dragged them to the side to smother the flames. They hadn't realized that the fire-retardant foam had disintegrated from the rain and, with so many people walking through the area, it had essentially become useless.

Ten firemen were injured and airlifted by helicopter to the hospital; three in serious condition with burns and smoke inhalation. Two would require skin grafts and months of medical attention. It would have been even worse if there hadn't been other firemen standing nearby holding nozzles aimed at the nose of the plane to quickly extinguish the blaze.

The fire was eventually contained and the bodies of the three pilots were pulled out of the demolished, charred plane. Guy and the two attendants were rushed to the hospital to be checked by medical personnel and then they had to report to the NTSB office to give an account of what had happened. When he and the flight attendants returned to the wreckage that sat on the golf course property to retrieve their personal items, no one wanted to go in. Since Guy was the youngest, he was volunteered by the others for the deed. He dutifully snatched up the attendants' purses and other personal items left onboard.

"I just stood there inside the cabin, staring at the cockpit and soaking in all the damage. It was hard to fathom. I heard that the pilots had passed when I was at the hospital. I'd already guessed they were gone, but it was still shocking to hear that people you knew and worked with had died. Two of them I knew quite well," says Guy. "I was so traumatized, I couldn't even remember my own telephone number and called a neighbor to reach my mother. A lot of people in our neighborhood saw the plane going down."

At the memorial service for the pilots, Guy saw the attendants for the first time since the accident and realized just how bruised and battered they were. As the mourners stood outside, a Mercer Airlines plane soared overhead in what Guy felt was a happenstance tribute to the perished pilots.

It was a short week at school, so Guy's mother let him stay home to collect himself before going back.

Understandably, Guy wanted nothing to do with planes for the rest of his life after his crash. He was somewhat of a celebrity at high school for a brief time but afterward was sure he'd never fly again.

His mother felt differently and encouraged him to "get back on that horse." She knew that he'd always loved flying and shouldn't walk away. A week later, Guy was back at his flying lessons. He was certified for his general aviation license when he was seventeen years old, although he would have to give it up nine months later for medical reasons. It took him another ten years to be truly comfortable flying, even on commercial airlines.

At the time, Guy was more preoccupied with where he would go to college and what he would do when he graduated than with the effects of the plane crash. Today, he reflects that it may have been fortunate that he was a teenager at the time and that, thanks to his youthful sense of invincibility, he didn't experience long-lasting traumatic aftereffects.

Guy feels blessed and proud to have been able to overcome the experience and return to flying, rather than living the rest of his life in fear. He says that people are often surprised that it was his mother who supported his return to flying—and he also credits her for teaching him how to move on with his life.

He points out that just because someone has a car accident, doesn't mean they stop driving. Flying is no different.

For many years, Guy thought he came through the accident unchanged. Later, however, he realized that it had given him a different perspective on managing challenges. As an example, he explains that when he was getting his two degrees there were times when he felt completely overwhelmed. But he would always tell himself that he didn't get this far to quit now—and pushed through.

"I skydived when I was fifty-four years old. I've lived a full life and was the least affected by the crash of all of the people onboard that plane. By the grace of God, I sat in the right seat. If I'd been in the cockpit jump seat like I had done before, I wouldn't be here," he says.

Not often, but every once in a while, Guy thinks about the crash. When the footage from the 9/11 attacks was replayed constantly on television, he couldn't help but think about what the passengers must have been going through as they plowed into the World Trade Center. There was also a crash in Colorado Springs in which a witness described seeing someone in the plane pounding on the window as it was going down. That brought back the terrible sense of claustrophobia Guy felt on his tragic day.

Obvious in hindsight, the NTSB investigation determined that the plane was not airworthy at the time of its flight. There was a fatigue crack in one of the blades in the No. 3 engine that had not been detected during an improperly performed engine overhaul. Airline companies are free to choose their own maintenance contractor but they are still responsible for ensuring that the work is being done in accordance with current manuals, equipment, and procedures—and Mercer Airlines' contractor was not doing so.

The accident resulted in a number of recommendations being made to the FAA, including how these types of propellers are inspected during overhauls and inspections.

TRIAL AND ERROR

According to a captain with the Los Angeles Fire Department, the DC-6 fire spurred revolutionary changes in fire service. For one, the Los Angeles Fire Department gave up using rotary saws and went back to using an ax to break open a fallen or damaged plane that might hold survivors.

Uniform shirts and pants in the mid-1970s were made from polyester. During daytime structure alarms, firemen wore turnout coats for protection but only polyester uniform pants covered their lower legs.

Polyester was popular at the time because it shed water and dried quickly—but it also melts. As the firemen at Van Nuys Airport tried cutting into the cockpit to rescue the pilots and the fire erupted, the polyester uniforms melted to the bodies of a number of firemen, increasing the severity of their burns.

About a year after the Mercer Airlines crash, polyester uniforms were replaced with sturdy cotton pants, a quantum leap in fire protection. Later, the department upgraded to Nomex®, which the fire captain calls "probably the best material ever developed for firefighting clothing." In the early 1990s, the Los Angeles Fire Department also changed its policies so that firefighting of all types would be done in full turnout pants and coats made from Nomex material.

TAIL HOOK

January 11, 1984
Piper T-Tail Lance
Basye, Virginia

Sandy Schenk. Photo credit unknown.

The NTSB report cited wing icing as the probable cause of the crash, but Sandy Schenk says that's not true. He'd had a similar issue with the same plane on a hot, summer day, although at the time he wrote it off to the air density at sea level.

To celebrate passing their certification exams for the American Board of Surgery, Sandy Schenk and another young surgeon from the University of Virginia Medical Center named John, along with two friends, flew to western Virginia to spend the day skiing. They got back to Sky Bryce Airport a little later than expected. The airport manager had gone for the day, turned off the runway lights, and left a message on the answering machine that the airport was closed. But Sandy never heard it.

Sky Bryce Airport is also not what most people imagine when they hear the word "airport." It's little more than a strip of tarmac about the width of a driveway smack dab in the middle of a field in rural Virginia.

The airstrip is relatively short and nestled in a valley between mountains, requiring a pilot to make a quick takeoff at a steeper angle than might be necessary at other airstrips. Sandy had flown in and out of this airport several times before, but not in the Piper T-Tail Lance he'd rented that morning in Charlottesville—and not in an aircraft loaded with four adults.

"I knew that the T-Tail Lance was an anemic climber, and I was having trouble seeing in the dark," he says.

The T-Tail design on the Lance resembles the tail of a mini-airliner and was only produced by Piper Aircraft for a few years. Most planes of this size are built with a standard vertical tail design and a horizontal stabilizer. Online pilot forums are filled with complaints about the plane being difficult to manage during takeoffs and in turbulence—and not suitable for short runways at all.

As it struggled to gain altitude, the tail of the plane hooked onto a building that was under construction, causing it to slide on its belly across the roof before launching off the other end and landing flat in a clearing. Sandy said hitting the building probably saved their lives because there was a heavily wooded area at the end of the runway that he doesn't believe would have been survivable had they crashed there.

John, who was in the front passenger seat, most likely hit his head on impact because his side of the windshield was completely gone when they came to a stop. He scrambled out of the hole in front of him.

In the T-Tail Lance, the windshield is constructed of two sections that are then put together. The pilot's side where Sandy sat was still completely intact after the crash, making his escape that way impossible.

Flames erupted, surrounding Sandy as he struggled with the door, but it was jammed. The door by his side also had double latches—one on the top and one on the side to keep the slightly curved door closed when in flight. The plane was full of fuel for their trip home and he knew what could happen.

"I had to get out quickly or I would be dead," says Sandy.

He relocked the latches and tried to reopen them again but had lost crucial seconds in the process. Sandy turned around and reached behind him to push the two other passengers in the back seats out of the plane through the hole where a door had once been, while John reached into the open side of the windshield and tried to pull Sandy from his shoulder harness.

An explosion erupted, blowing John away from the door and throwing him flat on his back.

John suffered some burns on his hands, a damaged tendon, and a cut on his face. The passengers in the back also had some blast burns from the explosion and one experienced chemical burns when she fell into a puddle of aviation fuel on the ground that had leaked from the damaged plane.

Severely injured, but still conscious, Sandy was strapped into an ambulance for the two-hour drive back to the University of Virginia.

"I was only seconds away from being able to walk away from the accident," observes Sandy.

He spent the next two years enduring two dozen restorative procedures. Scars from skin grafts still line his face, yet when

Sandy talks about the accident and his remarkable recovery, it is in a matter-of-fact sort of way.

"I lost parts of all of my ten fingers. I made a decision to do what surgery I could successfully and pursue that path. The biggest challenge was finding someone who would make custom-sized sterile gloves that I could use in the operating room," he adds.

As an academic surgeon at the University Medical Center, Sandy came to know the hospital's plastic surgeons after his own reconstructive surgeries. He was able to practice vascular surgery with his new hands and gloves by lengthening and shortening the femoral arteries on rats in the micro-vascular lab before he felt comfortable returning to surgery on people. Today, Sandy has a vibrant vascular surgery practice and is in great demand.

Sandy has learned how to adjust and to accept some of his limitations; he is unable to pick up a thin object like a dime from a flat surface. As a part of the teaching faculty, he is also not able to teach residents certain techniques, such as how to make a one-handed suture knot, because the sensation in his fingers is impaired.

Just before the accident, he had been appointed the head of trauma services at the medical center. Not wanting to be unable to handle any possible emergency that came through the door, Sandy resigned the position. He was also adamant about relearning his trade without the use of custom-made instruments. It took months of practice but he needed to be able to perform surgery with what was available and not be dependent on having instruments designed just for his particular hands.

At the same time, Sandy feels that his accident has given him insight that he didn't have before and that his patients appreciate not just his training at Duke University School of Medicine, but also that he takes the extra time to sit with them and review their options.

He remains highly respected in the medical community. In fact, a colleague at the University of Virginia was quoted as saying that Sandy is better than many surgeons that have all of their fingers.

About a year after the plane accident, Sandy visited the local airport—but the smell of aviation fuel sparked nightmarish flashbacks.

When he was ready to try flying again, he spoke with a flight instructor to explain what had happened. Expecting an easy-going reintroduction, Sandy thought the instructor was being a son-of-a-bitch when he threw the engine into idle and told Sandy to practice what he would do in the event of an engine failure.

His previous pilot training kicked in and Sandy maneuvered the plane toward a clearing until the instructor turned the engine back on as they neared an altitude of about one hundred feet above the ground.

The instructor explained in no uncertain terms that if Sandy wanted to be a pilot again, he didn't need sympathy. He had to

Photo credit Sandy Schenk.

be ready to handle an emergency. Otherwise, he had no business going back up in the air.

After a few moments, Sandy acknowledged that the instructor was right. He had to be prepared to handle any situation that arose during surgery—or while piloting a plane—if he wanted to resume the life he once had.

Sandy now owns his own plane and makes sure he is up-to-date on training and emergency procedures. Since his accident, the reported length of the Bryce airstrip was modified to 2,240 feet. This shorter length represents the usable amount of runway available to pilots, rather than the longer physical length of the runway previously reported that included areas that were unsuitable for planes. Sandy thinks this will prompt many pilots to think twice before using it.

"I'm more circumspect than I was in the past, and I always try to have a plan B when I fly. I had radar and a storm scope installed in my plane to avoid flying in poor weather conditions," Sandy says. "Life itself has risks but we have to learn to manage risk the best we can."

HONEYMOON HELL

September 13, 1982
McDonnell Douglas DC-10
Spantax Airlines
Malaga, Spain

Karen Stavert. Photo credit David Miller.

After the events of 9/11, Karen Stavert felt lost for nearly a year. The anniversary of her crash was just two days after the terrorist attack on the World Trade Center and it reopened the scab covering the pain and stress she'd worked so hard to overcome.

Nineteen years earlier, after two weeks of honeymooning in Spain, it was time for Karen and her husband to return home to Canada. Yet, for some unknown reason, just the thought of getting on the plane sent Karen into a tizzy, which was very

Photo credit Associated Press. Photo by Jay Boyarsky. Passengers of Spantax Airlines look back at smoke billowing from the smoldering wreckage of a charter DC-10 that crashed on takeoff from Malaga Airport.

unusual for her. She'd always loved flying and associated plane trips with the start of new and exciting adventures. But this time was different.

On the morning of their flight, Karen couldn't stop pacing their hotel room. She was lathered in sweat, in a complete state of distress. She told her husband she didn't want to get on the plane and pleaded with him to stay in Spain.

"We can't go. I don't want to go on this flight, I don't want to go home," she recalls telling him, feeling as though she could jump out of her own skin.

Her husband mistook her agitation for wanting to extend their vacation and told her not to be silly. They had to get back to work.

The charter flight operated by Spantax (a now defunct Spanish airline) was late getting to Malaga Airport on the southern coast of Spain where Karen and her husband were waiting. The flight had originated in Madrid and was making a quick stop in Malaga before heading to JFK Airport in New York City.

Karen watched with growing angst as luggage was loaded onto the plane, taken off, rearranged, and reloaded, as if there was a problem fitting all of the bags into the hold. It seemed to take forever and Karen was ready to explode.

"I felt helpless. There was nowhere to go, no one to talk to, I had this intense premonition," she says.

To board the plane, passengers walked across the tarmac and up a set of metal stairs. Karen wanted to scream and run but she kept marching forward, nearly comatose with the terrifying sense that something really horrible was about to happen. The only thing she could do to keep her mind off the turmoil boiling inside was to bury herself in a book.

To Karen, the plane seemed loaded beyond capacity. Overhead compartments were crammed full and she watched as people banged and shoved against the doors to force them shut. This was before airlines had begun restricting the amount of carry-on luggage per person, and many of the passengers had packages and bags stacked under their feet and on their laps.

As the aircraft picked up speed for takeoff and its nose lifted off the ground, Karen heard a loud *thud* and the plane began shuddering violently. The next thing she knew, the brakes were screeching and everyone onboard was screaming hysterically.

Unbeknownst to the pilot, the strong vibrations were likely pieces of tread detaching from the nose wheel. Fearing that he was about to lose control of the plane, the pilot decided to abort takeoff. There was only 4,250 feet of runway left and the DC-10 was moving at 126 miles per hour.

According to the investigation, the pilot's decision to abort the takeoff was considered "reasonable" given the circumstances and the limited time he had to make a decision, even though it was not in accordance with standard operating procedures. At the time, pilots were trained in the event of engine failure but not when other components of the plane (such as a wheel) failed.

The plane sailed off the runway and bashed into a low concrete building, causing one of the engines to detach. It broke through metal fencing and careened across the Torremolinos Highway, hitting several cars. Much of the right wing broke off after hitting farming equipment on the other side of the road. The aircraft had just been filled for the overseas trip and fuel started sloshing onto the ground as the plane skittered to a stop on a rock embankment.

Inside, luggage and packages flew around the cabin and rained down on the passengers. Lighting panels broke loose from the ceiling and crashed down as well.

Karen was sitting at a window on the left side, about a dozen rows behind the wing with her husband and a teenage girl. When the plane finally came to a stop, packages and debris littered the aisles, making it nearly impossible to move.

Someone pushed seats down to create a platform so that everyone could climb over the impassable aisles. The plane was filling with smoke and Karen heard her husband yelling. Karen was momentarily paralyzed, convinced that she was going to die and that there was nothing she could do except surrender. It was surreal, as though she knew a disaster was going to happen, and it was unfolding before her like a scene from a blockbuster movie.

Karen jolted out of her stupor as she felt something deep inside her scream, *NOOOO*. She wanted to live and needed to get out. Her husband, along with everyone else in the plane, was frantically pushing toward the exit as flames engulfed the side of the plane.

One man started pounding on the window, desperately trying to break through so he could escape. Other passengers screamed at him to stop because there were flames licking at the window that would certainly travel inside if the window was shattered.

Karen pulled her shirt over her nose and mouth to protect herself from the heavy, black smoke that was billowing through the cabin. Some of the emergency exits were not opening or were blocked by fire. Karen was shoved flat against one of the jammed exit doors from the crush of people behind her who were trying to escape.

The investigation into the crash found that both exits in the first section of the plane were opened, but only the door on the left side of the plane in the second section was opened due to the fire on the other side. However, a passenger opened the door on the right side anyway, and a few people managed to escape before fire destroyed the chute and flames and smoke shot inside the cabin where others were still struggling to get out.

In the third section, only the door on the left could be opened but soon that too became unusable as the fire burned uncontrollably. The crew attempted but was unable to open the two doors in the fourth section, which had most likely been damaged when it hit the concrete building. Evacuation was also hampered as passengers absurdly stopped to gather their packages and other belongings, rather than leave them behind. Fire and smoke roared into the back of the plane.

When the exit door Karen had been pushed against suddenly popped open, she and those behind her tumbled out like dominoes onto the wing.

The heat from the fire around them was overpowering but Karen couldn't move. She was trapped on the wing while her legs were still stuck inside the cabin. Other fallen passengers were sprawled on top of her, pinning her down.

"I don't know how long I was stuck, but it was long enough that I was severely burned. I could feel this hot, hot heat on my back and kept trying to pull my shirt down but it had burned away," Karen says.

Looking back toward the plane, she could see the horrified expressions on the faces of the people watching from inside. The pain was excruciating. She kicked until she freed her legs and rolled her entire body onto the wing.

Dazed, Karen called for her husband until she realized that he was already on the ground and so she jumped off the wing. Today, Karen is still haunted by the sight of an older woman nearby who was lying motionless on the scorching wing.

"I don't know what happened to her. I felt like I should have done something to help, but I didn't," she says.

The teenage girl sitting beside Karen during the aborted takeoff did not survive but her mother and sister who were sitting behind them did. Karen would later be told about an older woman with cancer who had been on the flight to take her last trip before the disease consumed her. She hadn't even bothered trying to get off the plane. Apparently, she'd had enough.

Karen's backside was badly burned. She was hustled onto a minibus as others were carried to the side of the road a safe distance from the plane. She moaned in pain as the wind from the open window whipped across her charred back. Each bump in the road aggravated her injuries. It was the longest nine-mile ride she'd ever experienced.

The bus finally arrived at Carlos Haya Hospital, a medical facility that looked more like a dingy, poorly lit garage. Karen was propped up in a wooden chair in the middle of a chipped bathtub. Using something that she described as garden hoses, the medical staff sprayed her with what she believes was a Betadine Solution and scrubbed her from head to toe.

"I didn't think the burns could hurt any more than they already did until then. I was screaming and don't remember much after that," says Karen. "The next thing I remember was waking up in the hospital with a tube down my nose."

During the ten days that she was in the Spanish hospital, an American friendship club helped her and many of the passengers contact their families back home. Initially, some of the news accounts had reported that all passengers had been killed.

Sucesos

El chárter de la muerte

La muerte viajaba a bordo de un DC-10 de Spantax

MAX Montalvo de veinte años, estudiante, sintió la muerte en su cuerpo. El DC-10 de Spantax, vuelo Málaga-Nueva York, se estrelló pocos segundos después de haber intentado despegar sin éxito. Al lado de Max, el suizo Hans Rudolf se palpó el cuerpo sobresaltadamente. Un grito: «¡Estoy bien, estoy vivo!». La novia que le esperaba en Nueva York ya ha podido abrazarle. Un matrimonio italiano volvía al tiempo al avión para rescatar a su hijo. Imposible. El avión era una bola de fuego. Un infierno. Dentro, más de medio centenar de cadáveres. Calcinados. Hechos carbón.

El comandante Pérez, jefe de vuelo, y su tripulación, 13 personas en total, habían llegado de Madrid en el DC-10 de Spantax para recoger unos pasajeros en el aeropuerto de Málaga y seguir viaje hacia Nueva York.

Todo en regla, ningún problema. Dos meses antes, el día 12 de julio, el mismo DC-10 había sufrido una avería en Málaga y fue llevado a reparar a Palma de Mallorca con sus 365 pasajeros a bordo. El vuelo inicialmente previsto, Palma-Madrid-Málaga-Nueva York, tuvo que ser rectificado sobre la marcha. La tripulación había comprobado que el motor izquierdo perdía aceite y los pasajeros comentaron que en el interior del avión olía a gasolina. Permanecieron en Málaga cuatro horas. Por fin, se despegó rumbo al aeropuerto mallorquín de Son San Juan. Allí se comunicó a los pasajeros que había «una fuga de aceite en los motores». El 13 de julio, martes, 36 ho-

Photo credit *Cambio 16*.

Of the 381 passengers and thirteen crew members aboard, forty-seven passengers and three crew members died. Eight people burned to death, while the others succumbed to smoke inhalation and heat-related injuries. More than a hundred others were hospitalized.

According to Karen, Carlos Haya Hospital had only two English-speaking employees: a young intern and a tiny orderly whom Karen estimates was no more than five-foot-five and one hundred pounds soaking wet. Both worked endlessly and dashed up and down the aisles, helping patients and interpreting for the medical staff.

The morphine didn't ease Karen's pain, but it did cause her to hallucinate. In her drug-induced delirium, she tore off her bandages and imagined that she had to submerge her wounds in water to put out the fire she thought was still scorching her skin.

"Morphine is horrible. I don't understand how people get addicted," she says. "The American evacuation team finally gave me different drugs that worked without the crazy side effects."

When it was time to leave Spain, Karen was still feeling no pain from all the medication she was taking. She wandered through the American military cargo plane that was racked with survivors of the Spantax crash on stretchers, entertaining some of the other survivors and their families with her babble, and was allowed to sit in the cockpit with the pilots as they flew to Fort Dix in central New Jersey. The Canadian Air Force then transported her and other Canadian nationals back to their home country.

Karen spent the next five weeks in a Toronto hospital. Eighty percent of her back was covered in third-degree burns and infection crawled through her wounds. In addition to the scars on her arms and back, skin grafts left what now resemble racing stripes on the back of her legs and the outline of her jeans pocket is permanently tattooed on her right buttock, a daily reminder of the burning plane's searing heat.

Karen had struggled with poor self-esteem for most of her life. When her mother told Karen that she was glad her daughter's face hadn't been burnt, Karen got angry.

"My back was the only part of my body I was comfortable with. Now, the only part I liked was gone," she says.

In the hospital, Karen felt safe. Nothing more could hurt her there. Outside, back in reality, it was a different story.

Once she was released from the hospital, she rode in a moving car for the first time and the events of the accident flooded back. She was terrified and wouldn't let any driver exceed a speed of thirty miles per hour.

Karen's fears and despair didn't subside in the aftermath; in fact, they escalated. Her anxiety ratcheted up anytime attorneys called asking questions about the crash. Karen started to drink heavily and more regularly to numb herself. When she and her husband went to New York for depositions, they traveled by train, thinking it would be safe. As Karen looked out her window, she felt fear and panic swell in her chest when she saw a burned-out shell of a train abandoned next to the track.

"It struck me then that it doesn't matter where you are or what you do. When it's time to go, it's just time to go," she says.

Over the next few years, Karen felt more and more out of control. One morning, after a night of particularly heavy drinking, she regained consciousness on the living room floor and heard a voice within her say, "You're almost there . . ." It was a huge wake-up call and she realized that the alcohol was slowly killing her.

Karen called the forensic psychiatrist in New York whom she'd been working with. The doctor ordered her not to drink and came to Toronto the next day to meet with her. Karen, in turn, took a train to New York for a follow-up visit and the doctor convinced her to enter rehab where she stayed for twenty-eight days. She transferred to the Hazelden addiction treatment center in Minnesota for another twenty-eight days before entering a halfway house.

Karen has been sober ever since.

When she retells the story of the Spantax crash, it's with a mixture of detachment and suppressed panic that leaves a slight quiver in her voice.

"It's almost like the crash happened to someone else, and in a lot of ways it did. I'm a very different person today," she says. "I choose happiness as much as I possibly can, rather than waiting for it to come to me. I used to be angry and unhappy, a real whiner and complainer. I didn't like the old me."

Karen continues to fly, mostly because there's too much of the world left for her to see that's not accessible by train. She's more comfortable in cars than planes but traveling on a bumpy road still takes her back to the accident.

To allay her fears, Karen relies on two alternative programs, Emotional Freedom Techniques (EFT) and Z-Point. Both have made an enormous difference in her life—they help her keep herself centered whenever she feels anxiety creeping up her spine.

Karen is now a glass and jewelry artist and teacher, amongst other things. Part of her healing process was learning how to use a torch and working with fire to make jewelry from glass.

"I needed to know what the cause of my fear was before any technique could be successful. It took some time, but I realized that the problem was rooted in the feeling that I had no control over my safety. Having to rely on someone else when I'm a passenger in a plane or car is hard," Karen says.

Although Karen learned her lessons of transformation in a traumatic way, she remains grateful. Today, she considers the scars on her body to be symbolic of her survival—not unlike a phoenix rising from the ashes.

KISMET

May 20, 1995
Cessna 172
Sechelt, BC, Canada

Lindy Philip. Photo credit Benjamin Philip.

As told by coauthor Lindy Philip.

BEEP . . . BEEP . . . BEEP . . . BEEP . . . BEEP!

In seconds, my thoughts skittered from *What the heck is happening?* to *Holy shit, we're in trouble!* The alarms coming from the front of the plane sent a paralyzing shiver of fear and confusion down my spine.

My life had been a bit chaotic recently, but this was beyond anything I could have imagined. Soon after getting married, I had felt a void in my life. It saddened me and I couldn't help feeling that our marriage wasn't complete.

I had even asked my sister, "Is this all there is?"

And then, one fine spring day, everything changed. I was meeting a friend for a drink after work. She was late and the group at a table nearby invited me to join them since I'd been sitting alone for so long.

That's how I met Ben. From the moment I looked at him, I felt a deep connection like I'd never felt with anyone else. There was an instant sense of familiarity and I knew he felt it, too. It was both exciting and quite confusing at the same time. I had been married for a little more than a year and had just met this charming man who lit up my world.

Our friendship flourished and I was reaching a point where I had to make a decision. I couldn't continue like this; it wasn't fair to my husband. My feelings toward my newfound friend (or shall I say soul mate) were growing immensely. I struggled with trying to decide if I should throw myself forward with Ben or try to work things through with my husband. I did care about the man I had married, but it was a different kind of love.

Eventually, I moved out and got an apartment by myself. I needed my own space to sort through the flurry of bewildered emotions.

Soon after, Ben and I decided to take a long weekend trip to the Sunshine Coast just north of Vancouver. It had been about eight months since we'd met and our relationship had blossomed into something meaningful. We felt so connected but there was considerable baggage lurking in the background. For me, the apprehension was creating a gaping hole in my chest that made me feel both ill and excited. Girlish flutters kept me up at night.

I hoped that my getaway adventure with Ben would put everything in perspective and help us determine our future. It was time to figure out if this was the real thing. I pleaded with myself, *Please help me find my way with this person. Give me a sign of where to go with this.*

When we arrived at Boundary Bay Airport, it was one of those magnificent mid-spring days. The sky was a sharp, brilliant blue without the slightest wisp of a cloud. A light breeze off the ocean gently caressed the land.

Despite the pristine conditions, my nerves jangled not only in anticipation of the weekend but at the thought of having to climb into a frigging little plane. A friend and work colleague, Ron, had recently gotten his pilot's license. Ron needed to log some flying hours and offered to give us a lift to the Sunshine Coast, an easy twenty-minute flight.

Breathe, breathe, breathe, I kept reminding myself, as the three of us climbed into the rented Cessna 172. My throat went dry and an eerie fluttering crawled inside my gut. Ben, on the other hand, was calm and relaxed, the kind of person who has no qualms about flying at all. I tried to drag my thoughts back to the upcoming weekend and my future, rather than ponder the absurdity of floating in the air in a small, misshapen metal box. But my exercise in positive thinking didn't work very well. It wasn't the first time I'd been in a small plane, but even the thought of riding in one had always caused my heart to skip a beat. This ride was no different.

Once in the air, we were floating and bobbling slightly, as small planes do, the loud whine of the single-prop plane making conversation nearly impossible. The blue-green Pacific Ocean waters sparkled as though littered with thousands of diamonds. It was such a glorious and surreal experience looking down at the beauty of the land from the back seat of the plane. I gazed out the tiny window, feeling a slight pang of nervousness, but tried to stay focused on the scenic view outside the window.

The welcoming sight of a small landing strip in a green field surrounded by hills appeared before us through the cockpit glass. Ben sat shotgun and I began fiddling as we neared the landing strip. From the back where I sat, all seemed fine as we started to descend and I waited to hear the slight thud and squeak of the wheels stroking the pavement, signaling the beginning of our magical weekend.

But something wasn't right. The plane was going too fast and it swooped past the runway. Ron was fighting with the controls, trying to get the plane back into the air.

Ben would later comment that, even as someone with no experience flying a plane, he knew the Cessna was not lining up properly with the landing strip. It was then that he became keenly aware of just how fragile the plane really was.

BEEP . . . BEEP . . . BEEP . . . BEEP . . . BEEP!

The plane's emergency alarms began screaming their warnings. It was disorienting and unsettling. Ron shouted into the radio, "MAYDAY! MAYDAY! Down draft, down draft. Wind shear. Can't get any elevation."

The nose of the plane tipped downward, heading straight toward a thick stand of pine trees on the side of a steep hill. The sky disappeared.

As I gaped at the fast-approaching trees, all I could do was whimper and choke out the words, "This is it." Ben wrapped his arm around the back of his seat and took my hand in his. I sank down into the crash position with my head down low, hugging my knees.

We plowed into the large pines and I felt myself let go, accepting the inevitable. We were going down. Fast.

The next moments were serene. A sort of slow motion, floating feeling overcame me, a sense of acceptance as I came to

Photo credit photographytips.com.

Photo credit photographytips.com.

terms with the fact that we were crashing; and there was nothing I could do about it at all, except hang on.

The wing clipped the top of a huge pine tree and the plane lurched down in a swift, dizzying ninety-degree swing. The branches lashed and snapped against the plane, tearing it to shreds while also rocking it like a giant cradle. We slammed to a stop. The seat belt garroted my waist and legs as the plane tried to catapult us out. All went black.

I don't think I was unconscious for long but have no way of knowing. When I came to, still dazed, I looked around at the wreck and embraced myself. I couldn't believe I was still alive. This overwhelming, yet momentary, sense of relief slipped away and my heart starting pounding when I realized that I was squashed in the back of a crashed plane. As I tried to breathe, white bolts of pain coursed through my body and I nearly passed out. The interior of the plane began to squeeze inward, sucking the air out of my lungs. I wanted out!

Ron and Ben were unconscious.

I shouted at them, "Wake up! Wake up! We need to get the fuck out! Get out!"

They started to groan, moving slightly. Knowing that there was a good chance that the plane could blow up at any moment, we crawled out, stumbling into the darkening forest. We dragged ourselves a distance away from what we were sure would be an imminent explosion, adrenaline overcoming the pain of broken bones. Blood streaked down both men's faces, their expressions a mix of complete horror and overwhelming relief before they collapsed on the forest floor.

Stupefied, Ron kept muttering, "I'm so sorry, I'm so sorry . . ."

"Dear God, Ron," I said, "we're all alive."

Ben sat dully, in complete shock, staring blankly at some point in the distance, unable to respond to anything.

It quickly became apparent that we needed to be found soon, before it got dark. Spending a night in the dense forest after surviving a crash was too much to even imagine. My breathing became ragged as I tried to gulp down the terror crawling up my spine at the thought of being stranded.

Thwack, thwack, thwack.

We could hear a rescue helicopter circling far above but there was no way they could see us from that distance through the thick pine-branch canopy.

Ron mumbled very slowly and was barely audible over the pain, "Go to the plane and get some flares."

Thwack, thwack, thwack.

I stood up and nearly keeled over. My entire midsection felt shredded after the impact that had thrown me against the seat belt. Walking was nearly impossible but I was still in better condition than Ben or Ron. With no strength left in my core, I struggled to lift each leg with my hands and place one in front of the other, the only way I was able to walk through the uneven terrain back to the plane.

I gingerly felt my way through the plane but couldn't find any flares. Instead, I grabbed my purse and camera (always the opportunist) and started snapping pictures of the wreckage.

Suddenly, as though from a scene in a movie, a mountain biker came charging through the woods. Expecting the worst, I'm sure, he stared at me with a baffled expression and no doubt thought, *Wow, you were just in a plane crash and here you are taking pictures. Seriously?*

Shortly after our discovery by the biker, more than thirty rescue workers, drenched in sweat and breathing hard, crashed through the branches and swarmed toward us. They moved quickly and started debating how to get us out; it was obvious that we couldn't do it ourselves.

There was talk of a helicopter airlift. The blood drained from my head and the scene around me started to swirl from the thought of being lifted through the thick tangle of pine branches in a stretcher connected to a helicopter by only a few strands of rope.

"Are you kidding me?" I said aloud. Not for me. Not even now. There was no way I was going to lie in a swinging basket dangling from a helicopter, even though I knew that, ultimately, I'd have to go along with whatever they thought best.

Fortunately for me, the rescuers opted for a ground rescue since the terrain was rugged and it would have been dangerous to attempt an airlift. The forest was soon filled with the deafening shrill of chainsaws leveling trees to clear a pathway as the evening's falling darkness complicated the efforts. The rescuers proceeded to put us on stretchers but first wanted to cut our shirts so they could examine our injuries with the least amount of movement. I distinctly remember Ben protesting; he didn't want them to cut his favorite shirt. *Is that really so important right now*? I thought. Ben came to his senses and relented.

Hours later, the rescuers finally got us out of the forest with great effort and to resounding applause. Ron was airlifted to Vancouver; his injuries and broken bones were the most severe. Ben and I were taken by ambulance to the local hospital in Sechelt.

During the ride, rivulets of sweat poured off the emergency technician who had just hauled us out of the woods. Despite everything that had happened, I squirmed, completely grossed out by the sweat that was dribbling off his face and onto mine as he secured my oxygen mask in place. It's strange, the things you remember during a disaster. Despite that, I was, of course, ever so thankful.

For the next three days, Ben and I were in the intensive care unit. The staff put us in the same room with our beds opposite

Photo credit photographytips.com.

Photo credit photographytips.com.

each other and propped up our pillows so we could see one another. We were quite the small-town celebrities. Complete strangers came to visit the people who had managed to survive a plane crash in their community.

It took at least six months for our broken ribs and soft tissue damage to heal, and we were advised by our physicians to rest and spend the summer at the beach. We convalesced at the aptly named Wreck Beach, which we had flown over just before the crash.

Our plane crash cemented our relationship and Ben and I decided that our journey together was definitely meant to be. I received my answer and divorced with no contention. In fact, my ex-husband gave us his well wishes. Looking back at my near-death experience, I have no regrets. I lived out my worst fear and I felt like I was saved for a few reasons—to carry on and be a mother.

But that doesn't mean I didn't struggle after the crash. I was in a fog for at least three months. I could not remember meeting people and often became disoriented. However, my healing process was hastened as I became more connected spiritually. I would sit in the woods by my favorite tree and meditate, feeling my wounds wash out as I communed with nature. I discovered through my healing (and through the impact of crashing to the ground) something else about myself. Despite having never been interested in art before, a creative part of me emerged and flourished.

My life changed its course and I married my soul mate, Ben. In my first marriage, we had agreed not to have children, but after meeting Ben and nearly dying, it was obvious to me that the universe had bigger plans—and that involved motherhood. I needed to be a mom and I am ever so grateful for being given that chance to raise two beautiful young men.

It is interesting how people often comment that the accident was bad luck. I see it differently. In fact, I was incredibly lucky to have survived such an ordeal.

I still travel, of course, and because we live on an island, traveling means flying. My kids love turbulence and feel that it is like a great and exciting amusement ride, especially when they ride on smaller island-commuter planes. I, on the other hand, do not share this feeling—not in the slightest.

When I was on a flight coming home from Mexico, I was sitting next to a woman who was terrified of flying. She needed comforting, and it felt good to be the one holding her hand, talking her through it and assuring her that everything would be fine. A friend who was traveling with me marveled that I chose not to mention my plane crash experience to my seatmate. I felt that it was unnecessary; it would have probably frightened her even more.

Now, any time I need to fly, I go through a consistent ritual. It involves drinking a Caesar or two (the Canadian version of a Bloody Mary) to take the edge off. Then, once in my seat, I use a Japanese relaxation technique called reiki. I am always concocting an exit plan, sitting on the aisle and in the back of the plane nearest to the door because I've heard that is one of the safer parts of the plane in the event of an emergency. People have mentioned that they want to travel with me because they feel the chance of a second crash is pretty slim.

Recently, I read one of my entries in an old writing journal from before the crash and was a little shocked at one particular paragraph—the passage mentioned a dream I'd had about being in a plane crash.

It gave me goose bumps to think that I had had this dream years before the crash happened.

Ben, on the other hand, felt exhilarated after surviving the crash and has tempted fate on several occasions by trying his

Photo credit Benjamin Philip.

hand at skydiving and riding in a small Challenger jet. He swears that death never crossed his mind that May afternoon in 1995. I can't quite say the same thing.

The two of us have adopted an anonymous quote as our mantra to remind ourselves that each day is a gift. These words are forever in our hearts:

The past is history
The future a mystery
The moment is a gift and that is why it is called
The present.

STEVE'S DISCOVERY

Steve, the mountain biker who first discovered Lindy, Ben, and Ron after they had crashed into the hillside, shares his story:

That day, I was out riding my mountain bike along the Chapman Creek ravine and was surprised to come across a police car on the old logging road trail. The police officer, a friend of mine, asked if I had seen or heard anything unusual.

I hadn't. He told me a plane had gone down just after takeoff from the nearby airport and he was the first one to respond. Since I was familiar with the area, he asked me to take a look around until other responders arrived. I immediately headed down the ravine to find out if I could see anything.

It wasn't long—maybe five or ten minutes—before I heard muffled moaning drifting out of the forest. This was a good sign; the thought of finding dead and mangled bodies strewn on the ground was already making me extremely nervous.

Amazingly, the plane was in one piece, but totally destroyed. And while the occupants of the plane appeared to be injured and very shaken up, they were alive. The pilot was the worst off; he couldn't move.

I assured them that help would soon be there and took off on my bike to where the emergency responders were located. I told them where to find the plane and hung around for a while, but honestly, my part was done. The pros had taken over and I went on my way.

Any time I think about that day, it still amazes me that a little plane could go down in heavy trees like that and everyone was still able to walk away.

Perhaps not literally walk away—but we know what he meant.

Photo credit Stuart Robinson.

DANCING IN CIRCLES

July 11, 1978
Piper Cherokee
Manukau Head,
New Zealand

David Shenkin. Photo credit David Shenkin.

Never in a million years could eleven-year-old David Shenkin guess what lay ahead as he clambered into a borrowed Piper Cherokee plane and headed off to Mount Hutt, a ski resort on the South Island of New Zealand, with his older cousin and his father, an experienced amateur pilot, at the controls.

The trip was supposed to be a skiing adventure but the weather didn't cooperate. The three of them were cooped up in the ski lodge for the entire nine days. The mountain, nicknamed Mount Shut, remained shrouded in clouds the entire time they were there and never opened for skiing.

"There was no television, smart phones, or friends for me to play with at the lodge. I was bored to hell and the only thing I could do for entertainment was annoy my cousin. Even though I wanted to get out of there, the weather was so bad that I had serious misgivings about getting back into that plane," says David, thirty-seven years later.

A sense of dread churned in the pit of his stomach, giving way to adolescent fantasies of sabotaging the plane so they wouldn't have to leave. His father, however, had his own ideas about flying in this weather. David adored his father and describes him as a charismatic attorney who was unshakeable whenever he made up his mind to do something. And he certainly wasn't about to let a storm like this keep him from leaving the mountain, especially after nine days of being caged in a ski lodge.

The mood was mixed. Happy to end a vacation that had kept them cooped up for nine days, but wary of the weather, the trio flew into the heavy and oppressive blankets of clouds. As they entered Cook Strait, the strip of water that separates the North Island from the South Island, fierce headwinds barreled down the steep, jagged cliffs and battered the small, four-seat plane. Skis, boots, and luggage went flying around the cabin, further terrifying the two boys.

David's father decided to land in the city of Wanganui on the west coast of the North Island to refuel and get an updated weather report. The boys whispered to each other about their hopes of staying put but they were both too young at the time and had no say in the decision. David's father strode out of the airport office like he'd just received a favorable decision in court and announced that the three of them would be continuing to Auckland despite the churning winds. Beside himself with fear,

David talked his cousin into taking the copilot seat that was normally reserved for him.

"I wanted to get out of the front seat. I had this terrible feeling that we were going to crash," David says.

It didn't take long for his father to realize that trying to continue the trip up the west coast was a mistake. They were pitching wildly in the air and had to get back to the ground. They tried to land in New Plymouth four or five times but the crosswinds gusted so strongly that the plane nearly slammed into the ground on one of the passes and then was almost tossed into the control tower on another attempt.

"I can still remember seeing the look of horror on the guy's face in the tower," David says. "We tried several small inland airfields but they refused to let us land because they didn't have emergency equipment on hand and the runway wasn't long enough if our landing didn't go well."

The airport at Taupo in the center of the North Island finally gave David's father permission to land, saying that the winds were manageable but that the plane would have to get there before nightfall. David's father was an experienced private pilot but he was not instrument trained to fly at night or during poor visibility.

Hearing the news, a flood of relief swept over David. They would be able to land before sundown; he now had hope that they might live. But then he realized that the aircraft was slowly turning away from Taupo and back toward the west coast.

"I was a chatterbox when I was a kid and my father always told me to keep my mouth shut in emergency situations," says David. "I still don't know if he turned deliberately or not and this has been a source of incredible pain for me for years."

According to David, his father would have found it quite inconvenient to be stuck in the small town of Taupo and would have to rent a car to drive hours back to Auckland. He was scheduled to be in court the next morning and the two boys had school. Today, David tries to give him the benefit of the doubt about why he turned away from Taupo, but it's not always easy.

David's heart sank as he heard his father radio the airport at Taupo, saying that he'd gotten disoriented and was back on the coast and needed advice on how to get to Auckland International Airport. He wonders if his father was really that hardwired to mislead the air traffic controller. According to David, the small plane was capable of going only about one hundred knots and was now fighting against eighty-knot head winds (Category One cyclone winds) that slowed them to a crawl.

Just when David thought it couldn't get any worse, the instrument lights blinked on. His father seemed to go pale and David was strangled by fear as the air traffic controllers advised his father to fly parallel to the white line of frothing waves crashing on the coast below them.

By this time, the plane was running dangerously low on gas. The Auckland airport closed its runways and got emergency equipment ready in preparation for the four-seater's landing. Air traffic controllers told them to turn right at the lighthouse that marked a gap in the rugged coastline in order to make their approach to the airport.

David saw the flashing light first and his father made a turn to where his son pointed.

"What we were told by air traffic control was incorrect," David explains. "There was no lighthouse, just a flashing beacon indicating a weather station approximately two kilometers short of the Manukau Harbor entrance."

When they turned, the plane was exposed to eighty-knot crosswinds that pushed them farther southeast—and their fate was sealed. They were now being pounded by turbulence and were flying only a few hundred feet over very rough countryside. The plane tore down the cliffs and was tossed through the air like a brittle leaf somersaulting from its fragile hold on a sturdy branch. His father cursed as he wrestled to regain control of the lightweight plane as it bucked and lurched. The air traffic controller urged him to correct his course.

"My cousin's head was rolling from side to side. There is only so much fear you can take," David says, clearing his throat and collecting himself before he could continue. "I must have passed out too because I don't remember the impact. The fear was suffocating."

David would later learn that the right wing of the plane was torn off when it grazed the top of a ridge, hit a shed, crashed through a fence, and flipped into a gully. Had the plane been just six inches higher, he believes they would have made it.

When air traffic control lost contact with the plane, they called the local weather station, which alerted farmers on the rugged Awhitu Peninsula that a small plane may have gone down nearby. Despite the howling winds, one farmer emerged from his shelter to search his property in the storm. Hunched over his motorbike, he was barely able to see through the rain and the fog even with his headlight on.

The farmer drove across his property three times before he noticed a deep gouge in the turf. As he approached, he saw that his fence had been knocked down—and the remains of a plane lay motionless just beyond. He rushed back to his house and called for help.

When David regained consciousness, the wind was shrieking through the silenced plane broken only by a loud banging of a door somewhere. His seat belt was broken. With his head throbbing and one eye swollen shut, he dragged himself out through a small cargo hatch by leaning down to feel the ground as he inched out into the pitch black. He heard someone

Photo credit Stuart Robinson.

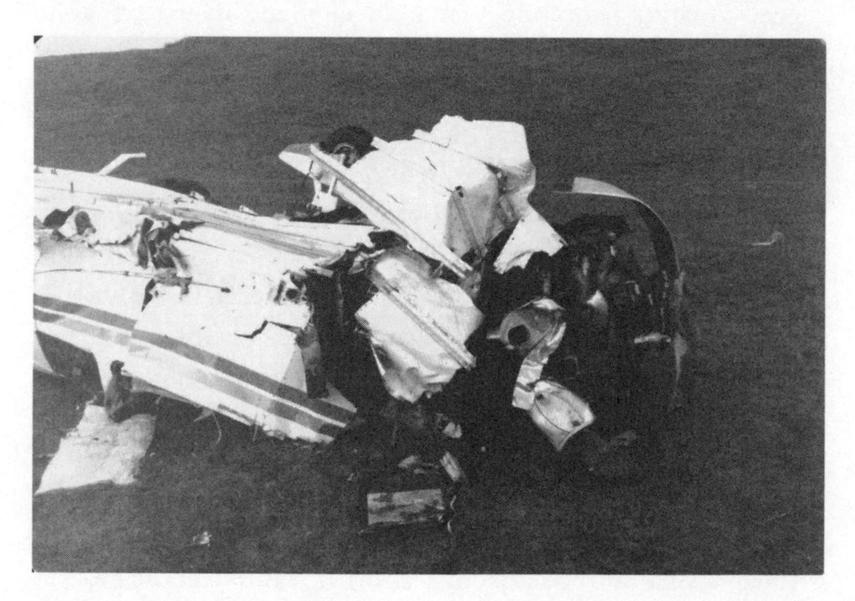

Photo credit Stuart Robinson.

groaning. When he tried to pull his cousin out, he let out a terrifying moan. His cousin looked like a bag of broken bones and he was losing a lot of blood.

David stumbled around the plane, frantic to find his father. He doesn't remember if he did, and doesn't want to remember either. His father died on impact. He was thirty-seven years old. There was nothing David could have done.

Exhausted, David curled up on the ground under the wing for protection from the weather. Wrapped in his heavy jacket, he hoped sleep would take him away from this nightmare. The next moment, he saw brilliant headlights that reflected in the roiling mist. He used his last burst of energy to run towards the light, waving frantically and yelling.

"I had a head injury and my face looked like it had been beaten with a baseball bat," says David. "I couldn't see through my left eye."

Telephone calls ricocheted from house to house. By sheer luck, a party was in progress nearby and one of the guests was a doctor who had recently outfitted his station wagon as a modified ambulance complete with blood plasma. That stormy night was the first time the doctor had taken his homemade emergency vehicle out. He immediately set off to the crash site, where he gave David's cousin enough blood plasma to keep him alive until he could get him to the hospital. It was nothing short of a miracle that his cousin had lived; according to the hospital team he was minutes from death, nearly dry from loss of blood, and had to be resuscitated repeatedly on the operating table.

David was loaded into a Land Rover and rushed to the ambulance waiting on the rough, unpaved road. During the long trip to the hospital, he faded in and out of consciousness.

By then, news of the accident had reached his mother, who was waiting at the hospital when he arrived. When the hospital staff asked her to identify the unconscious boy, David was so

Photo credit Stuart Robinson.

disfigured that all she could tell them was, "It looks like his hair, but I can't tell if it's him."

After being X-rayed, a nurse covered David's eye with a patch so he would look, in her words, "beautiful" for his mother, who was just being let into the room. When she walked in, he started vomiting the blood he'd accidentally swallowed. It was horrific for everyone.

David kept asking for his father but no one would tell him what had happened. Yet, he knew in his heart that his dad was gone. Finally, a young female doctor came into the hospital room and told him flatly that his father was dead, before abruptly walking out. There were no formalities and there was no one to talk to. Life was expected to go on, even though everything had just turned upside down. Years later, David finally pieced together why no one had visited him in the hospital in the days following the accident. His relatives and friends had been at the funeral, which would have been in keeping with the Jewish tradition of burying their loved ones soon after their passing.

David's father had been the center of his life. His parents were in the middle of a divorce and had divided the children between themselves three years earlier. David lived with his father, while his brother and sister lived with his mother. Now that his father was gone, David moved back with his mother—and that's when the trouble started.

The first day he returned to school with plaster still partially covering his face, David was whipped with a strap for not finishing his homework. Before his death, his father had had an ongoing battle with David's headmaster. His father knew David was a difficult child who was wired for mischief and who had trouble sitting still and concentrating. Now that his dad was gone, David was sure the headmaster (whom he describes as hostile, sadistic, and a repressed homosexual) grabbed the opportunity to retaliate and ordered the teachers to use corporal punishment to keep him in line.

Instead of becoming obedient, the nearly daily beatings drove David to rebel against authority figures at school, at home, and with the police. At twelve years old, he was taking family cars out for drives, shoplifting, and stealing anything that wasn't nailed down. At fourteen, marijuana became his highly illegal savior and held back the demons consuming him since the crash. It didn't take long for David to be kicked out and shipped off to boarding school, where he was bullied for being the only city kid and the only Jew.

He was further ostracized because he couldn't play sports due to his facial reconstruction and a knee injury he had sustained after riding a stolen motorcycle into the back of a parked car a few years earlier. David admits that he probably made a tactical error in thinking that the best way to blend in was to be arrogant, which didn't win him any points with his schoolmates either. When he dropped out of boarding school, his life continued its downward spiral. At one particularly low point, he found himself with a gun pressed against his head by a neo-Nazi.

David landed in Sydney with twelve dollars in his pocket when he was seventeen years old, and scraped by working odd jobs and living in hostels. When the hip-hop nightclub scene kicked off in Sydney, he inherited some money and fell into the world of ecstasy dance parties. The money was quickly swallowed by cars and women as chaos reigned in his life, and he sank further and further into a deep hole of depression masked by drugs.

Looking for answers to the unanswerable, David moved from Australia back to New Zealand, then to England and Israel, among other places. He became religious and then non-religious, but drugs seemed to be the only thing that squelched his anxiety and gave him peace.

While living at a yeshiva in Israel, he used pot to still himself and study the Torah, but he couldn't absorb what he was reading whenever he was high. Eventually, he was asked to leave. David spent an increasing amount of time with his Israeli war veteran friends whom he says were as "fucked up" as he was. He attended a few drug addiction meetings and was clean for over three years, but then had a drink and quickly became a raging alcoholic. It didn't take long for him to slip back into smoking pot again.

On a visit home to New Zealand, he stopped by to see an old pal who introduced him to crystal meth.

"I joked with my friend that I was going to use it until I was either broke or in rehab. My mate laughed and told me later that he knew it would come true," David recalls.

It wasn't until David was toeing the ledge of a ten-story building in Bangkok, a breath away from allowing himself to lean forward and fall, that he started coming to his senses. As he stood there, a silent scream roared in his chest. He finally came to the realization that drugs were taking more than they were giving and his mind flashed back to an incident more than ten years earlier when he witnessed a car lose control and flip on the highway. David ran to the wreck and helped pull two passengers out of the car, just before it burst into flames.

"It finally dawned on me that if I could save a couple of guys from a car accident, I must be worth saving myself," David says.

For the next three and a half months, David was in an inpatient addiction rehabilitation program. There, he finally grieved as he had never grieved before, not just for his father but also for the good friends he'd lost tragically to a bombing in Jerusalem, a car accident, cancer, and murder at a mass shooting.

"What do you do with this kind of grief? The prospect of facing it was always too painful to think of," says David. "It all came out. I cried and cried and owe them my life."

Rehab also helped David come to terms with the gut-wrenching sense that he was the one who should have died in the plane crash that had haunted him for twenty-six years. Survivor's guilt had fueled his drug abuse and reckless behavior along with the suicidal urge to complete the task himself.

David feels that his life would have been different if his father had been there to help him grow up. It's disappointing that his dad wasn't able to watch his children and grandchildren become adults.

David's father might have been an influential man at the top of his game, but he was also reckless and born seemingly without fear. His father had always said that he wouldn't live to see his fortieth birthday, and had even said to David, about a month before the crash when they pulled into their driveway, "If we're ever in an accident and one of us has to die, I want it to be me."

City Lawyer Killed as Plane Crashes At Heads: Two Hurt

A prominent Auckland lawyer was killed when a light plane crashed on the southern tip of the Manukau Heads last night.

He was Mr Brian Kevin Shenkin, of Riddell Rd, Glendowie.

Two other people in the plane were seriously injured. One was Mr Shenkin's son, David, aged 11, and the other Douglas Ross, aged 17, of Mission Bay.

The injured youths were taken to Middlemore Hospital in separate ambulances. After a tortuous journey in heavy rain the ambulances arrived at the hospital about 10.30 pm.

Douglas Ross was rushed to surgery with head injuries. His legs were broken in several places. David Shenkin has head and chest injuries.

The three were returning from the Canterbury ski field at Mt Hutt when the plane crashed into a ridge, smashed through a fence, dipped under high-tension power lines and clipped a hayshed, tearing one wing off. The plane then cartwheeled and landed on its roof.

It is believed Mr Shenkin was flying the plane.

Late last night the two youths were in a serious condition in Middlemore Hospital.

Wreckage

Wreckage from the crash was spread over several hundred yards and it was difficult to recognise any of it as being from a plane, a policeman said at the scene.

The plane was an Auckland Aero Club single-engine Cherokee.

It was thought to have been battling strong easterly winds on an approach run to Mangere after dark.

First on the scene was Mr [illegible]

He and his wife, Joyce, were watching television when they had a telephone call from the operator at the Manukau signal station, Mr Frank Rowe.

Mr Robinson rode his motorcycle to have a quick look around his rugged farm. About threequarters of a mile from his home he found "a bit of white stuff."

Upside Down

On looking further he found the earth ploughed out of land near an old airstrip on the farm.

"Then I could see the plane upside down," he said.

He dashed back to his home and raised search and rescue authorities before returning to the scene with neighbours.

Mr Robinson said: "When we got out there you couldn't see a thing. It was foggy like pea soup."

Registrar

Ambulance men had to use Land-Rovers to reach the wrecked aircraft because of the ruggedness of the country.

Mr Shenkin graduated from Auckland University and as a final-year law student represented New Zealand in a moot with Australian universities in 1962.

He worked for the legal firm of Rudd, Garland and

Horrocks. With another lawyer he won the Sir Robert Stout Shield, contested by students of the Auckland Law School.

In later years he was best known as an expert in divorce law.

In June, 1975, he was appointed registrar of private investigators and security guards.

His appointment followed the introduction of the Private Investigators and Security Guards Act.

Mr Shenkin

Article from *The New Zealand Herald*, July 12, 1975.

As a result of the investigation into the crash and a series of other recent air accidents, the Auckland coroner asked the Ministry of Transport to consider giving air traffic controllers the authority to ground limited-license pilots to keep them from flying in poor conditions.

The Chief Inspector of Air Accidents was later quoted in the local newspaper as saying that pilots operating under visual flight rules "must also have sufficient self-discipline to make appropriate action to avoid being trapped by low cloud or bad visibility."

After a month of sobriety, David rode his motorcycle to the field where the Piper Cherokee had gone down. When the same farmer who had found them came out of his shed, David started crying. The man knew without being told that David was the boy who had survived the accident on his property all those years before, and embraced him.

During his decline, David had absented himself from his family, perhaps to protect them, but also because he felt ashamed. His family consisted largely of university professors and successful professionals. For a long time, he felt cheated as his siblings went on with their careers and family lives, but now he understands that they suffered too. They had lost a father, an ex-husband, and a son. Everyone was affected.

When he finally spoke to his family to make amends for his years of destructive behavior and explained that it had been rooted in the stress he lived with every day since the crash, his uncle said, "That was such a long time ago; we thought you were over it." But for David, the trauma from the accident had never stopped. He had obsessed over it every day of his life for more than two decades.

After rehab, David finished high school and returned to Israel for some time. He had always wanted to marry a Jewish woman but joked that there were only two or three in New Zealand and they were all related to him.

"My wife actually discovered me singing 'Born to Run' at a karaoke bar in Israel. I tried to break the relationship off, worried that it wouldn't work with all the baggage I carry, but she wouldn't hear of it," he says.

David and his wife started a family and he returned to school to become a therapist in order to help others with addiction issues. Since then, he has also changed his perspective on the crash and believes that after all the pain he endured, he's better equipped to handle dark issues and be there for others when they need it.

He rarely talks about the crash, but when he does, the emotion is still raw in his voice. Even when he was a desperately lonely eleven-year-old boy, he never wanted to be viewed as the poor kid who lost his father that way. When people used to tell him that he must be destined for something special, having survived all that he had, David would reply that he used to buy into that notion, but it all still felt like a crushing responsibility. He had even tried to confront his past when he was in his twenties and signed up for flight school, but he would break out in psychosomatic coughing fits—and he couldn't get off drugs at the time either.

Today, on commercial flights, David is hyperaware of everything that happens on the aircraft and is extremely nervous during takeoff and landing. He doesn't really ever feel comfortable until they are at a high-cruising altitude.

"I'm not afraid of death; I don't know why," he says. "I've lived a colorful life, but being clean and sober and happy for the first time in a long time is tremendous. For so many years, I was deeply depressed, suicidal, and drug addicted. I lived through horrible shit. Now, I live one day at a time."

GROUNDED

December 28, 1978
McDonnell Douglas
DC-8
United Airlines 173
Portland, Oregon

Aimee Connor. Photo credit Peter Ford.

The first clue that something was wrong on United Airlines Flight 173 was a loud *bang* and an unsettling shock wave that reverberated through the DC-8 as it approached Portland International Airport.

Captain McBroom's intercom announcement explained to the passengers that they were having a little trouble with the landing gear but that everything was okay—they should relax. Passenger Aimee Connor said the flight had been ordinary up to this point. She was just a teenager at the time and believed what she was being told.

Photo credit Associated Press. An aerial view of United Airlines DC-8 which crashed in a residential neighborhood near the Portland International Airport in Portland, Oregon.

But the captain wasn't being entirely truthful. The three men in the cockpit were extremely concerned. The thump was highly unusual and the plane had yawed slightly to the right when it happened. In addition, the only indicator light that had come on was for the nose gear. There was no way for them to know from the instrument panel that the main landing gear was actually down and locked in position for landing. Instead, the head pilot assumed the worst.

To Aimee, no one seemed particularly worried, even when a member of the flight crew came into the cabin with a flashlight, looking through one of the windows to check if a mechanical tab had popped up on the wing, indicating that the landing gear was engaged.

Around the same time, an off-duty pilot who was a passenger onboard was heard on the cockpit recorder as saying, "Less than three weeks, three weeks to retirement. You better get me outta here." The pilot told him not to worry, but also suggested that he put his coat on for easy identification and protection in case it got "hot."

The captain came back on the intercom to tell everyone that the airport was putting out foam for a belly landing and that it would be controlled and safe. Aimee, an invincible seventeen-year-old at the time, thought that meant that soft foam was being laid out on the runway, not understanding that the pilot was referring to fire-retardant foam used to minimize an impending disaster.

Aimee was returning to boarding school after a family gathering with her grandparents in New Mexico to celebrate their fiftieth wedding anniversary. She and her parents first flew together to Dallas, where her parents then continued onto Minneapolis while she made a connection for what should have been a two-and-a-half-hour trip to Oregon.

The plane made it to Portland in the allotted time but circled the airport for an hour while the captain tried to puzzle out the landing gear issue.

The flight attendants went row-by-row, making sure the passengers knew how to get into the brace position and which emergency exits they should use when they got to the ground. Some passengers were moved around so that stronger men would be able to open the exit doors and help others get off if necessary.

The pilot was giving updates every few minutes, and Aimee felt reassured. A women sitting next to her, however, was very frightened. To calm her down, the head flight attendant told her earnestly, "If I thought for one moment that we were in any real danger, I would be the first one to jump out of this plane with a parachute."

Meanwhile, there seemed to be a serious misalignment between the seriousness of the fuel situation and how the flight crew was behaving. Much of the cockpit recording captured the captain obsessing about the landing gear and not paying attention to warnings from the other members of the flight crew or even the blinking warning lights on the fuel gauges that go off only when the levels get so low that the engines take in air.

Even when Captain McBroom called the United Airlines office in San Francisco to get advice, in the recording he sounded almost casual, "We got about 165 people onboard and we . . . want to . . . take our time and get ready and then we'll go. It's clear as a bell and no problem."

The pilot finally announced that they would be landing in five minutes. Then, all of the cabin lights went out. And the plane went eerily silent.

"Other people said they remembered the engines going out at that time too, but I only remember someone yelling to get into the brace position," says Aimee.

When the first officer told the captain that they'd lost an engine, the captain seemed surprised and asked "Why?" even though the first officer and the flight engineer had tried several times to tell him that fuel levels were dangerously low. Then, all of the engines lost power.

The fuel gauges in the DC-8 seemed to be contributing to the confusion but the investigation would later determine that they were working fine. Regardless, the NTSB report pointed out that a commercial airplane should not fly with anything less than forty-five minutes of reserve fuel at any time.

The captain called in a mayday. The last radio transmission to the air traffic controller was: ". . . the engines are flaming out. We're going down. We're not going to be able to make the airport."

He aimed the plane toward a darkened area, hoping that it was a field or some other uninhabited area.

Still unaware of what was really happening, Aimee thought she felt the plane touchdown but the bobbling sensation was actually the plane rubbing the tops of the trees in a suburban neighborhood about six miles southeast of the airport.

Then mayhem ensued.

The cabin was filled with the thunder of metal and wood ripping to shreds as the ninety-ton plane plowed a scar in the earth 1,500 feet long and 130 feet across in the winter evening darkness. With no gas left in the plane, there was no fire or explosion, just the unforgettable sound of complete and instantaneous devastation that echoed through Aimee's bones.

Tree branches tore off the left wing and high-tension electrical lines were ripped down before the plane plowed into and flattened one home, skidded across a busy road, and demolished a second house. The first house was vacant and for sale, and the person living in the second house had just gone out for dinner. No one on the ground was killed and the plane miraculously missed a nearby apartment building and did not collide with any cars on the road.

Photo credit Associated Press. The cockpit of United Airlines DC-8.

The cockpit broke off from the first-class section and crumpled under the body of the plane, leaving what one witness would describe as "an area no bigger than a dining room table for six." An investigator described the front of the plane as looking like it had been peeled back like a banana.

The floor buckled and Aimee's leg was pinned between seats. Two men helped pry them apart and she limped through a hole where the left wing had been. All Aimee wanted to do was get out of the plane.

She recalls that her fellow passengers were calm but very confused because they didn't know what to do. Many of them would later say that they don't even remember how they got off the plane, just that the next thing they knew, they found themselves standing on the ground in stunned disbelief. Emergency vehicles hadn't yet arrived and residents in the neighborhood, jarred by the noise and shock waves from the crash, poured out of their houses, astonished to see the plane separated into pieces where a house once stood.

"I started worrying about my mother. She'd recently had brain surgery after an aneurysm," says Aimee. "I thought if she heard about the crash in the news, her blood pressure would skyrocket and she would have another one."

Aimee doesn't remember much, only hobbling with her battered ankle into one of the nearby houses and asking to use the phone to call her family in Minneapolis. When she explained to her father that there had been trouble with the landing gear, he assumed that she was calling from a pay phone at the airport and that everything was okay. In all fairness, Aimee admits that he might not have been taking her seriously at first because she had been a bit of a drama queen in her teens.

"I could tell he didn't really believe me or understand what I was trying to say. But I knew they'd heard my voice and would know that I was okay when they learned what had happened on the news," Aimee says. "I'm glad I made that call."

When she dialed home again a couple hours later, news reports were saying there were only some survivors and her parents were beside themselves with worry.

Ten people died—eight passengers, the kind flight attendant who'd tried to comfort Aimee's seatmate, and the flight engineer. Twenty-three others were seriously injured.

Aimee gladly accepted a neighbor's offer to come into her house for coffee and some warmth but she was collected by a police cruiser and brought to a local church where the airline was doing a head count and having the survivors checked by medical staff. At the church, a television reporter cornered Aimee for an interview before she was transported to the hospital to have her ankle X-rayed.

The NTSB report concluded that the probable cause of the accident was:

> . . . failure of the captain to monitor the aircraft's fuel state properly. This resulted in fuel exhaustion to all engines. His inattention resulted from preoccupation with a landing gear malfunction and preparations for a possible landing emergency.
>
> Contributing to the accident was the failure of the other two flight crew members either to fully comprehend the criticality of the fuel state or to successfully communicate their concern to the captain.

The investigation would also later determine that corrosion had caused the landing gear on the right side to slam down rather than lower in a controlled manner. Because it came down so fast, it caused one side of the plane to have more drag as well as the slight yaw described by the cockpit crew. It wasn't until the landing gear on the other side was in position that balance was restored. The sudden drop also disabled the circuit that controlled the light indicators in the cockpit.

Aimee's original plan had been to stay in Portland with some friends before catching a short flight to school in neighboring Washington State the next day. Her friends saw her being interviewed on television and realized they'd forgotten to pick her up at the airport. To gain access to the hospital, one of her friends pretended to be her cousin to get through security and up to her hospital room.

"The nurse thought I was in shock, because when she said your cousin is here, I just looked at her blankly," says Aimee. "It wasn't until she said 'your cousin Keith' that I started to put it together. It was such a crazy day."

Aimee stayed with her friends in Portland for a week before returning to the airport to catch a flight on a small regional carrier to Wenatchee, Washington. That would be the only flight after the crash that Aimee would enjoy. Even after the fourteen-seat plane hit some turbulence that clearly disturbed her seatmate and other passengers, she was totally at peace, convinced that there was no way a crash could possibly happen again. The woman next to Aimee looked at her perplexed and asked, "Aren't you scared? Did you not know there was just a large crash in Portland?" All Aimee could do was smile serenely.

Back at school, the enormity of what had just happened started to sink in. Aimee knew that the crash was a big deal—she'd been interviewed by television newscaster Walter Cronkite, after all—but was still surprised that everyone in her small community was in shock and relieved to see her. The first night, she stood at the podium at dinner to tell her story and then slept for thirty-six hours straight, something she had never done before or has done since.

Aimee said everyone around her seemed to think it was important to get back on the proverbial horse. However, a few months later during a flight to Minneapolis to visit family for spring break, she couldn't stop weeping and wailing and gnashing her teeth. She thought she was having a nervous breakdown.

Aimee flew seventeen times between 1978 and 1985, trying biofeedback, large quantities of alcohol, pharmaceuticals, self-hypnosis, and other techniques to calm herself—but nothing worked. With each flight, Aimee only became more panic stricken and loud in her terror. After each trip, it took weeks for her to rebound.

Before boarding a plane, she made it a practice to tell flight attendants about her history, which usually earned her a free gin and tonic. But the last time Aimee flew, she was completely out of control. The flight attendant had to sit next to her the entire trip. Once she landed, Aimee hugged her mother, kissed the ground, and vowed never to get on a plane again.

"It was so embarrassing. I needed so much one-on-one attention that the attendant wasn't able to do her job. I felt sorry for the people sitting around me; it had to be very uncomfortable for them," says Aimee.

Nothing in Aimee's life would ever be the same. After she graduated high school, Aimee briefly moved back home and became a Quaker, attracted to the small spiritual community that was cognizant of issues surrounding peace and social justice. But she quickly realized that Minnesota was not right for her. Aimee had always enjoyed the Pacific Northwest and had friends who were moving back, so she decided to join them. Ironically, she now lives about five miles from the site where United Flight 173 crashed on that cold December evening.

It took her about ten years before she gave up trying to be a "normal" flyer and sought post-traumatic stress disorder counseling. Only then was Aimee able to accept the fact that her anxiety about flying was not unreasonable for someone who had survived a crash. She was not a victim but a survivor who hadn't fully healed.

From that point forward, Aimee became an Amtrak traveler. She was also fortunate to have met someone who had suffered his own loss and who has a good understanding of Aimee's emotional triggers. He's also a train aficionado, with a large train set in their basement. They are a match on many levels.

In 1998, Aimee organized a twentieth reunion of the survivors, responders, and people who lived in the neighborhood where they'd crashed by enlisting the help of a local newspaper columnist who did an interview and ran articles about the reunion. The idea spread like wildfire, and Aimee installed a second phone line in her home just to field calls about the potluck reunion. Hundreds showed up, and Aimee reunited with her seatmate. One brazen survivor admitted to marching up to the United Airlines desk at her local airport with a copy of an article about the reunion. She insisted that United fly her to Portland free of charge. The manager agreed and she got a standing ovation for sharing her story.

When Flight 173's pilot reluctantly came to the reunion with two of his daughters, he also received a standing ovation. He was blamed for the crash and lost his pilot's license, but a number of survivors thought he was a hero for getting the plane down with so few fatalities. The pilot's daughter, whom Aimee still stays in touch with, shared that her father felt deep remorse over the accident that resulted in ten people losing their lives.

"I thought he was the bravest person for coming to the reunion. He made a bad decision because he thought he had more fuel and he was ruined after the crash. I never thought he was a hero, but no one in his right mind would have crashed an airliner on purpose. He never intended for this to happen and was obviously a broken man," explains Aimee. "There was a lot of forgiveness that night."

Aimee says that while she wouldn't wish a plane crash on anyone, she feels that she's a better person today because of it. That is not to say that she wouldn't have reached the same place at some point but the trauma of the crash helped her gain fresh insight into herself and other aspects of her life and the world around her.

"I was given the opportunity to forgive and learned some important lessons early in life. It helped me reevaluate how I was approaching challenges and figure out better ways of handling them," she adds.

At the same time, the crash continues to affect Aimee. She worked as a hospice nurse for years and feels that her struggles made her more compassionate in the care she gave others, until lingering post-traumatic stress swallowed her whole after the 9/11 attacks in New York. Aimee lost her ability to read for a period of time and she became agoraphobic. Leaving the house to make the three-block trip to the grocery store was nearly impossible and driving became more difficult.

Her life is quiet now and she tries to instill peace in her daily life.

COCKPIT RESOURCE MANAGEMENT

The 1978 Portland crash changed the dynamics of the airline industry forever for the better and was highlighted on a National Geographic television series segment named *Fatal Fixation* that included interviews with Aimee and an actress playing her as a teenager in the reenactment.

The head pilot of Flight 173 was very experienced. He'd been with United Airlines for twenty-seven years and had logged more than 27,000 hours in the air by the time of the accident, 5,500 of which as a DC-8 captain. In addition, the first officer and flight engineer had thousands of flying hours between them. It was hard for anyone to understand how the crash could have happened, how such an experienced pilot could have circled within sight of a major airport for an hour without being able to land safely.

In the 1970s, the culture in a cockpit was driven by the hierarchical authority of the captain, many of whom were described as cocky and arrogant as one would imagine an ex-military Top Gun pilot might be. Challenging a captain's reasoning about something as basic as fuel would have been extremely difficult, even when a preoccupation with a landing gear issue was obviously blinding him to the impending emergency.

Aviation expert Dr. Alan Diehl served on the crash investigation team as the NTSB Human Factors Group cochairman. He recommended the implementation of a new program that was being developed by NASA called Cockpit Resource Management (CRM) training. CRM would improve teamwork and collective decision-making, especially at times of crises when no one person could possibly be cognizant of all potential threats to the safety of a flight.

It was not a hard sell. The Ames Research Center at NASA was already researching training concepts to reduce human error and improve safety. According to their research, poor communication and teamwork in decision-making processes was a contributing factor in over 70 percent of all airline accidents. Two prime examples around this time were the Eastern Airlines crash in the Everglades in 1972 that killed 101 people and the collision of KLM and Pan Am jets on a runway in the Canary Islands in 1977 that killed 585 people.

Over the years, a number of programs have evolved to improve communications on civilian and military planes and helicopters. Studies have shown that these techniques have significantly reduced accident rates by as much as 81 percent.

The underlying concepts of CRM have also been adopted in some countries to help overcome cultural barriers of excessive politeness and/or reluctance to challenge authority that could shroud a dangerous situation.

CRM was put to the test on July 19, 1989. The collaborative environment in the cockpit of United 232 was credited for avoiding a total loss of life (185 of the 296 people onboard survived) when the plane crash-landed in Sioux City, Iowa, without the use of conventional control systems.

In CRM-style programs, participants acquire tools to help prevent common errors that can arise from other distractions. Leadership, teamwork, and defined responsibilities are emphasized along with other skills such as stress management and coping strategies, as well as recognizing and controlling counterproductive attitudes and behavior styles.

Versions of this training have since been incorporated into programs for flight attendants, maintenance crews, and other roles in the aviation industry as well as in unrelated industries such as healthcare, firefighting, and executive business management. Flight simulators have been used to improve the interpersonal skills among high-performing teams by optimizing collaboration and individual performance.

One survey in the healthcare industry demonstrated a 78 percent reduction in surgical errors; 99 percent reduction of wrong site surgeries; 83 percent increase in employee satisfaction and patient safety awareness; 67 percent decrease in surgical infections; 48 percent decrease in waste of disposable goods in the operating room; 89 percent reduction in nursing turnover; and 63 percent increase in operating room turnaround times.

A COMPASSIONATE CON

When it was announced that the plane was going to make an unexpected landing, two passengers sitting in the back of United 173 volunteered to man a rear emergency exit so they could help others get out after it landed. No one knew at the time, but one of them was a convicted robber who was being escorted by Captain Roger Seed back to the Oregon State Penitentiary. When the plane stopped tearing through the neighborhood, it became wedged between trees and the slide on one of the emergency doors couldn't be extended.

Uncuffed, Kim Edward Campbell helped lower passengers from the emergency door seven to eight feet below where Captain Seed was standing on the ground. Kim then pulled the officer back into the plane and the two checked the entire cabin to make sure no one was still onboard. Captain Seed went out the door and, when he turned around, Kim had slipped through another exit into the darkness.

Eleven months prior, Kim had walked away from an Oregon work camp with only fourteen months left on a four-year sentence. He was recaptured in Denver and was on his way back to jail in Oregon when the plane crashed.

News accounts quote Captain Seed as saying Kim was an enormous help and could have escaped in the chaos when the plane first crashed, but didn't. Many of the passengers wanted to thank him for his help, having no idea that he was a prison escapee, although that fact probably would not have made a difference in how grateful they felt for his help.

Photo owned by Tejas Sreedhar.

A LIFELONG PASSION

February 27, 2014
Piper Tomahawk
Oxford, New Zealand

Tejas Sreedhar. Photo credit Matthew Lang.

Flying had always been a passion for Tejas Sreedhar and he worked hard to become a respected flight instructor, a career he intended to pursue for life. That dream, however, was seriously sidelined in February 2014.

Tejas had worked as an instructor at the International Aviation Academy of New Zealand in Christchurch for more than seven years and says he had a reputation for being overly cautious and pedantic about safety.

Which is why he finds being a plane crash survivor so ironic.

On the day of his accident, Tejas was flying a Piper Tomahawk with an Indonesian student sent to the Academy by Air Fast Indonesia to train for a commercial pilot's license. According to Tejas, the student was part of one of the first groups of candidates sent by Air Fast as a trial. It was a potentially large account for the New Zealand flight school—and two of the four initial candidates were entrusted to Tejas.

The Tomahawk is a two-seat plane originally designed by the Florida-based manufacturer Piper Aircraft, Inc. for flight training, touring, and personal use. It was introduced in 1978 and manufactured until 1982. The plane was designed to require a pilot to intervene to recover from a spin, rather than the plane self-correcting as other small planes do, to improve proficiency levels for student pilots in these situations. This feature was also largely responsible for the plane being dubbed the "Traumahawk."

According to Tejas, the student was doing very well overall with his training. But the flight school insisted that he make up some flight time that evening after hours, despite Tejas's objections that it was unnecessary.

The evening was perfect for flying. There was a light wind, perhaps five to ten knots and virtually no traffic in the usually busy air space due to the late hour. The sky was a brilliant blood orange. It was one of those magical evenings that occasionally descends on the east coast of the South Island of New Zealand during the summer.

Tejas and his student were flying for about an hour, doing standard training exercises that included a forced landing without power, which teaches a pilot how to respond in the event of engine failure. The exercise, which starts at about 3,500 feet to give the student ample time to locate a potential landing spot, simulates the steps needed to bring the aircraft to the ground safely.

"As an instructor, I've done this maneuver countless times under all kinds of weather conditions without ever having trouble," says Tejas. "During the exercise, the engine is idling. You don't actually turn it off but the plane behaves as though it has no power as you descend."

Tejas went on to explain that, rather than actually landing, the plane is powered back on after the student goes through the motions of an emergency landing. The instructor then reviews the descent and provides feedback to the student on how they handled the situation.

On the second forced-landing practice run, the student came in a little too high for Tejas's liking. If this had been a real emergency, the plane would have landed too late and run out of a place to land on the makeshift runway before it could safely come to a stop.

The plane climbed back up and Tejas suggested another attempt.

Again, the student slightly overshot the simulated landing. But as he switched to full power, the engine started coughing and sputtering. The Tomahawk was struggling to stay aloft and was losing altitude. The nose of the plane was at a steep incline in an effort to keep the plane in the air, but it was also dangerously close to stalling. Stalling is a procedure that is practiced regularly among those learning to fly and is very familiar to an experienced instructor like Tejas; however if a plane goes too far into a vertical position, it can lose all forward momentum and could drop like a rock.

Tejas went into overdrive in a desperate attempt to revive the engine. In the few seconds it took for him to quickly check the cause of the engine failure, they overshot the farmer's paddock that they'd used to practice emergency landings just minutes before. The next paddock wasn't an option. There were log piles and other obstacles scattered about. If they tried to land there, the plane would have crashed with a full tank of highly flammable fuel. Such a collision would certainly have been the end of them.

The plane continued to slide downward as they headed straight for an ominous row of tall pine trees that Tejas estimates stood about thirty to forty feet high.

"I thought about trying to fly parallel to the trees but the plane was going so slow at that point that it would have stalled if I tried a roll," Tejas says. "I was very aware that we were about to crash. The student had turned control of the plane over to me, and the only thing left to try was to balloon lift the plane over the tops of the trees."

With panic crawling in his chest at what was unfolding before his eyes, Tejas fought with the plane to keep it airborne, keenly aware that his actions over the next few seconds would be vital to their survival.

As they drew close to the trees, Tejas pushed the nose of the plane down sharply to put it in a harrowing dive and then used the added acceleration to pull back before striking the ground to generate a temporary lift up and over the trees.

Remarkably, the maneuver worked. But just as they were clearing the trees, the Tomahawk's right wing nicked a branch. The plane momentarily pivoted around the snagged wing and then plowed into a nosedive.

Tejas was electrified with fear as the ground came rushing toward them. The propeller started spinning from the downward acceleration, causing the engine to rev and squeal, a siren of the upcoming disaster.

As they crashed through the trees, in what Tejas described as feeling like slow motion, the cockpit was filled with the thunder of metal crunching and snapping. Then everything went still.

The entire front of the cockpit was ripped away but Tejas remained strapped in his seat and fully conscious. Unable to move, he was struck by the absurdity of the view in front of him. There was nothing but open air, the bright orange evening sunset, and a view of the same trees that had snatched them from the sky. The battered propeller lay useless on the ground a few feet in front of the mangled plane.

In agony, Tejas could see that his left leg was smashed and grotesquely twisted at an impossible angle. He and the Indonesian student gasped a few brave, but brief, words telling the other that they were alright. But neither one of them really was.

The Tomahawk was equipped with an automatic emergency location transmitter designed to send a signal in the event of a crash like this. But Tejas was unable to move and had no way to check if it had indeed activated as it should have, nor could he reach the controls to manually turn it on. There was no way for him to know if the transmitter had alerted anyone about their accident. They had crashed near Oxford, a small bucolic farming community of less than two thousand people. Who knew if anyone had even seen the plane go down, much less could direct rescue efforts toward them?

The smell of aviation fuel was overpowering and the risk of explosion real, but Tejas couldn't move. He was jammed in the destroyed plane with his shattered left leg. He was stuck in a death trap and couldn't escape. He started to feel sick from the pain and panic.

Other students from the school had crashed in the past and died. In most cases, there was little left of them and the images of their funerals with just a small box of ashes instead of a coffin flashed through Tejas's mind. He could not rid himself of the thought that this too was about to become his destiny.

The student managed to pass Tejas a cell phone and he called the 111 emergency services number for New Zealand. The

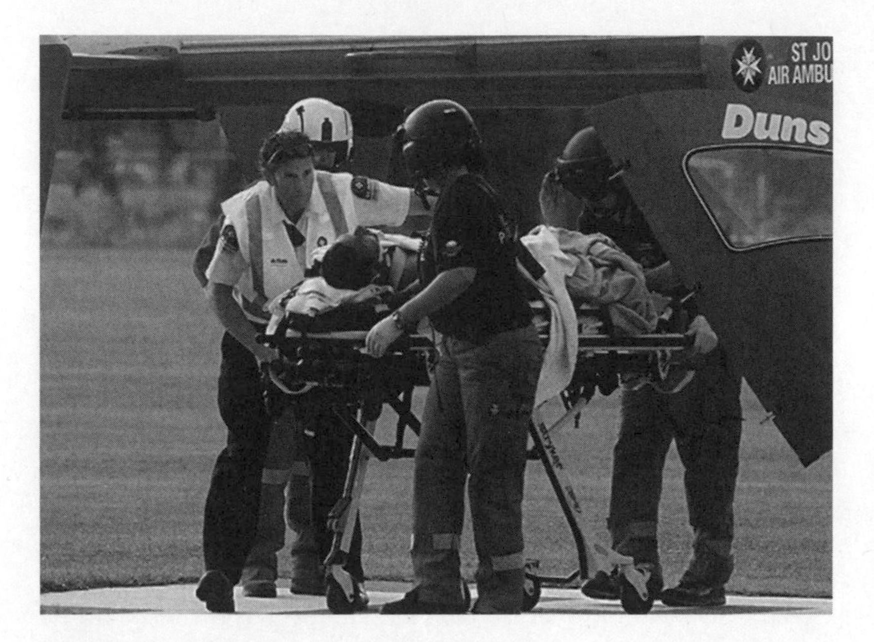

Photo owned by Tejas Sreedhar.

operator kept Tejas on the phone with repeated questions about his location.

Tejas couldn't turn to see the student who was now silent; he wasn't sure if he was even still alive. Tejas's stomach started bloating from internal bleeding. The hospital would later tell him that he had come within twenty minutes of dying.

When the helicopter arrived an hour later, they did not have the equipment needed to cut the pair out of the plane and had to fly back to retrieve their Jaws of Life. Tejas believes that since they were expecting to extract bodies—not living persons—from the site, they hadn't bothered to bring the life-saving equipment. A paramedic who stayed behind told Tejas to shut his eyes while she gave him something for the pain. The next thing he remembers is waking up in the hospital, and his next nightmare started.

For five weeks, Tejas underwent multiple surgeries. Two were needed to stop his bowel and spleen from bleeding. The bone of his left leg was crushed, leaving fragments that resembled pebbles below the skin. Papers were signed for an amputation of his left leg. He was only thirty-three years old. But, at the last minute, the surgeon decided to try to reconstruct the bone with metal.

Before he finally left the hospital, Tejas would have another four surgeries on his leg and a metal plate inserted into his arm to repair a break there. In addition, smaller procedures repaired a gash under his chin and his veins that had rejected the intravenous lines. The student, in the meantime, was treated for two broken ankles.

"The surgeon and medical staff did a tremendous job but the pain was excruciating," says Tejas. "I didn't know it was possible for a human being to tolerate pain of this magnitude. It changes how you feel about everything."

Even just moving the sheet on his bed caused Tejas to roar out in earth-shattering screams. Every nerve in his body was hypersensitive.

During his time in the hospital, management from the flight school visited Tejas frequently, wanting to talk about the accident—even when the hospital staff advised them that he should not receive visitors.

"People from the flight school were persistent and tried to convince me that it was not the engine that had failed but this was impossible for them to know. An investigation hadn't even been started yet. They were trying to manipulate my memories," says Tejas. "The school had never had someone survive a crash of this magnitude and they wanted to protect their reputation."

At the time of this interview, seven months after the accident, Tejas, who hobbles in an orthopedic "moon boot," knows that he will live with the consequences of the crash for the rest of his life. It's true that his injuries could have been worse, but that thought is not much of a consolation for him. The healing process is painfully slow and frustrating.

Reflecting back, Tejas says that he couldn't help but think about just how close he had come to dying.

"I know you can't live your life as though each day is your last but I've come to realize that a lot of things that used to worry me really aren't important. What's important is spending time with the right people," he observes.

The crash and post-crash experience helped Tejas realize just how fickle people can be. He couldn't count on some whom he had considered to be his friends for years, while other acquaintances came to his aid without being asked. There were times when he called friends in desperation, just needing someone to talk to, but he was frequently ignored.

"I have both lost and gained faith," he says. "I don't have time for people who are insincere or dishonest."

Dealing with his previous employer has been traumatic as well. Tejas is infuriated by what he calls their ridiculous attempts to blame him for the crash, which has added to the emotional toll of the accident.

"I'm usually such a positive person but the investigation has really gotten in the way of my recovery. My health should have been the most important goal at the time," he notes.

Tejas explained that the New Zealand Civil Aviation Authority (CAA) opted not to investigate his crash because there were no fatalities. While still in the hospital, he learned that the CAA was going to rely on the flight school to conduct their own investigation. The findings would then be submitted to the CAA for their approval, which Tejas equates to allowing an accused murderer to investigate the case against himself.

"Stamping that investigation report with a CAA seal of approval is a waste of time because my employer would never admit any fault. There is a lot riding on the result of the report. Air Fast was very concerned after the accident nearly killed one of the first students they had sent," says Tejas.

He wants to continue to be a flight instructor but resigned from his position at the Academy due to the mishandling of the accident. He plans to be very careful about who he works for in the future; he wants to be part of a group that believes in safety, honesty, and transparency. Flying has never been just a job for him. It was—and still is—his passion. But he's discouraged by what he sees as business interests interfering with safety.

After his accident, Tejas was told by a number of pilots that the Tomahawk he was flying that day was not up to snuff. He had not flown that particular plane much in the months leading up to the accident and others said that both students and instructors routinely avoided it. This claim was supported by the

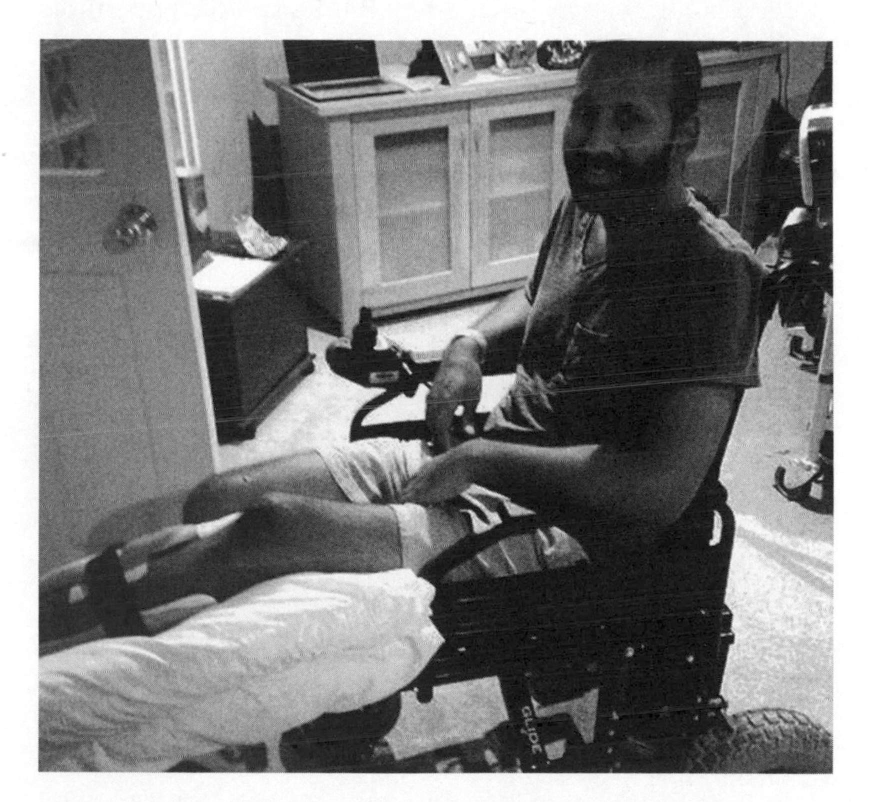

Photo owned by Tejas Sreedhar.

significantly lower hours flown on that plane when compared to all others of the same type at the flight school.

After the crash, the wreckage was hauled to a hangar at the Academy rather than to an independent location. Tejas is worried that if the report takes his or the Indonesian student's words out of context, it could impact his ability to work as an instructor.

He added that the CAA has an obligation to investigate crashes to protect the public and avoid future incidents. Tejas and the student both came dangerously close to dying and it pains him that the CAA would not investigate the accident independently. Finding a disinterested third party, Tejas acknowledged, is not easy in a small country like New Zealand where everyone in the aviation industry knows each other. But it's still the responsible thing to do.

"I've always preached the importance of honesty and safety in everything I do, and I tell students that if they damage a plane, they need to get over their embarrassment and report the incident for the safety of others," concludes Tejas. "Not making the cause of this plane's partial engine failure public could jeopardize other people's lives. I can't bear that thought."

AMAZING GRACE

June 1, 1999
McDonnell Douglas MD-82
American Airlines Flight 1420
Little Rock, Arkansas

Lisa Rowe. Photo credit Lisa Rowe.

Lisa had just spent a week in Colorado Springs visiting a dear friend and was traveling back home to Arkansas by herself. Colorado always held a special place in Lisa's heart. She had been stationed there for thirteen years with her husband and her daughter was born there as well. When her friend dropped her off at the airport, however, neither had any idea how Lisa's life was about to change forever.

Photo credit Associated Press. Photo by Mike Wintroath. Officials look over the wreckage of an American Airlines MD-82 jet in Little Rock, Arkansas, after it skidded off a runway while landing during a gusty hail storm, broke apart, and burst into flames.

Her flight was delayed. American Airlines offered passengers the option of taking a flight the next morning or waiting a few hours for the plane to arrive. Lisa's husband was starting a new job the next day and her teenage daughter had her first driver's education class scheduled. It was also her twentieth wedding anniversary and she wanted to be home, so Lisa opted to stick with her ticketed travel plans.

The weather was stormy for most of the trip to Dallas where Lisa was to catch her connection to Arkansas. She didn't fly very regularly and prayed for a safe flight throughout the bumpy ride.

"When we landed, the airport was chaotic. People were everywhere and I was running around like a crazy woman trying to make the second flight," explains Lisa, clutching the hand of a good friend as the conversation drags her back to that day.

The airlines announced another weather delay and for some reason Lisa's seat had been changed to a row closer to the front of the plane—an innocuous move that, upon hindsight, may have saved her life. She called her husband again to give him an update so he'd know when to pick her up at the airport in Little Rock. Some passengers were getting frustrated; others were chatting and playing cards to pass the time. As Lisa tells her story today, it pains her to recall casual conversations she overheard among fellow passengers who did not survive the day.

Boarding a plane for the second time that day, Lisa hustled to her seat and prepared herself for what should have been a quick forty-five minute hop home. She started praying again as soon as they took off into the air. She distinctly remembers the pilot telling the passengers to look out the window on the left at a "spectacular lightning show." Rather than being entertained, the faces of the passengers around Lisa seemed very tense; no one was amused by the pilot's suggestion. The flight seemed to get rougher by the second, if that was even possible.

As they prepared to land, the winds were blowing at twenty-eight knots and gusting up to forty-four knots, speeds associated with tropical storms. The air traffic controller had already issued two wind-shear alerts.

"I thought we were all going to die. I could see that the stewardess was scared too," says Lisa, her voice ragged with emotion.

As American Flight 1420 began its descent, bolts of lightning ricocheted off the clouds around them and an intense clap of thunder sent bone-chilling vibrations through the cabin. The plane pounded the ground and launched back into the air, crashed back down again and lifted up one last time as if in a feeble attempt to get airborne before thundering down for the last time.

Instead of the familiar grip to the tarmac that passengers feel with a forward motion against their seat belts, the plane began to hydroplane and fishtail on the slick runway. Lisa said everyone in the cabin sat in stunned silence at the surreal sights and sounds around them.

Engines shrieked and the entire plane began shaking violently, sending shock waves through every cell of Lisa's body as she gripped the armrest with all her strength. The plane carrying 139 passengers and six crew members never seemed to slow.

Lisa was sitting in seat 14A next to the window and was, for the first time, uneasy that the only thing protecting her from the nightmare unfolding outside were the sheets of metal of the plane's outer skin that now seemed so thin and inadequate.

Barreling down the runway, the out-of-control plane tore through tubes protruding from a metal frame structure that supported one of the runway's approach lighting systems. It crashed through a chain link security fence and passengers were tossed and thrashed about in their seats as the plane tumbled down a rock embankment into a swamp near the banks of the Arkansas River.

The deafening sound of metal twisting and tearing filled the cabin. As the plane lost power, the cabin was thrown into total darkness.

The collision with the lighting stanchions killed the captain instantly and caused the plane to break apart and ignite. Stanchions are usually designed to break apart on impact in the event of an accident, but these had been reinforced to keep them from tipping over because they were erected in the soft soil of a wetlands area.

The bins overhead broke loose and crashed down onto Lisa. Her heart ached as she watched two girls traveling alone cry out loud that they loved their mother and father.

The plane broke open like an egg. Rain and hail the size of quarters pounded through the jagged hole as thunder rocked the shattered mass of metal. In terrible pain, Lisa could barely move. Her seatmate helped her struggle through the hole in the roof to move away from the flames that were consuming the remains of the plane. She had to step over the lifeless body of a woman pinned down by metal.

The investigation would later determine that, in the rush to land before the thunderstorm worsened, the pilots made mistakes. They failed to engage the automatic braking and automatic spoiler systems necessary to stop the jet, especially important at night and in inclement weather. Consequently, there was minimal slowing of the plane, and when the captain applied too much reverse thrust on the engines, the aircraft spun out of control.

Standing on the top of the plane, Lisa was terrified that the lightening sizzling wildly in the sky above would strike the metal

Photo credit Associated Press. Photo by Mike Wintroath. The wreckage.

plates under her feet and electrocute them all. But it was pitch-black and she had no idea how far she'd have to jump off the plane or where she would land. As she glanced into the cockpit, she could see the nearly decapitated body of the pilot slumped in his chair, a sight that will haunt her forever.

Thick, acrid smoke and the sizzling sound of fire sent her skittering down the side of the plane. The only light came in bursts from the blinding white flashes tearing the sky apart, as rain and hail pelleted Lisa and the others stumbling forward as they tried to flee the burning plane.

"People were running everywhere. We couldn't see anything except when there was a flash of lightning, but we had to get away from the plane in case it blew up," Lisa says.

Unable to get around the aircraft and up the steep embankment, about two dozen survivors had no other choice but to escape through the flooded swamp. Pushing through the murky water, Lisa began to panic; she had never learned to swim and the water was already up to her chest. After surviving a plane crash, Lisa was now convinced that she was going to drown. She didn't know at the time that they were only about twenty-five feet from the strong currents of the Arkansas River that would have certainly swept her away.

The group staggered to a small island that rose slightly above the level of the water around them, and huddled together on the ground shaking from the cold, shock, and explosive racket of thunder.

"I couldn't get to where the others were. I was too hurt. I prayed and tried to sing 'Amazing Grace' with them," says Lisa. "Even today, I still don't like to be alone. I thought I was going to die alone on that little island."

After about an hour of waiting, wondering if they'd been forgotten, flashlights stabbed through the sheets of freezing rain and hail, signaling that rescue crews had finally reached them.

Lisa was having trouble breathing—her heart and lungs were bruised during the failed landing, but she somehow mustered her remaining strength to wade back across the swamp. Weak and in shock, she kept slipping under the water but was pulled back to the surface by a man walking behind her.

Responders thought she was having a heart attack and hurried her into a waiting ambulance. On the stretcher next to her, she could see the skin peeling off the severely burned arms and legs of a teenage girl.

Lisa passed out.

At the hospital, a chaplain leaned down and asked, "How are you?"

"I thought, *what a stupid question*, and answered 'I'm lovely, how are you?'" Lisa quips.

By this time, it was about two thirty in the morning, and no one knew where Lisa was. Her husband would be waiting at the airport, and her tenth-grade daughter would be sleeping soundly at home. Lisa's husband didn't have a cell phone, so she called a friend she knew would be home to tell her what happened. The friend contacted the airport to get word to her husband and then dashed to the family home to be with her daughter so she wouldn't find out about the crash from someone else. According to Lisa, American Airlines didn't tell any of the people waiting at the airport what was going on. These relatives and friends only heard about the accident when three busloads of ragged survivors showed up at the terminal.

"My daughter was scared to death when my friend came by. She thought someone was trying to break into the house; she didn't know what to do," says Lisa with the obvious pain any parent would have for causing their child distress.

According to Lisa, the hardest part of the whole ordeal was the healing process, not just from the physical injuries that kept her in the hospital for a week, but from the ensuing survivor's guilt that would come to affect everything she tried to do. There was so much to process and deal with and she found herself sometimes lashing out in anger at friends and family for no reason.

She survived, but five passengers perished as a result of injuries sustained during the crash, and five more died from smoke inhalation or heat-related injuries. Of the remaining souls onboard, more than 80 percent were injured. Many of them suffered terribly, including a man who sat in the seat originally assigned to Lisa who broke his neck.

A faculty member from nearby Ouachita Baptist University was traveling with the school's choir, coming home from a European tour with two of his children and twenty-five students. His severely burned, fourteen-year-old daughter was transported in Lisa's ambulance and died two weeks later while doctors were amputating her leg.

By all accounts, the captain of the plane was a seasoned pilot. However, at the time of the attempted landing, the winds exceeded the crosswind limit in these types of poor conditions for the MD-82. The landing should have been abandoned and the plane redirected to another airport.

Lisa still keeps a light on when she sleeps, can't go places by herself, and is keenly aware of where all exit doors are no matter where she is. Small incidents have a way of upsetting her, and she is obsessed with the fear of being alone when she dies. But even when she is difficult to be around and unlovable, her family and friends have stuck by her side, for which she is eternally grateful.

For years, she went to therapy several times a week to learn how to live with her survivor's guilt, and admits that she never thought post-traumatic stress disorder was a real diagnosis until she had to deal with it herself.

"I'm still upset with the pilot. His poor choices affected all of our lives," says Lisa. "It's also hard for me to understand why an old lady like me survived and these young people didn't. I struggled for almost fifteen years but had to forgive in order to be free to heal. I needed God to get my thoughts and feelings together. I hope and pray never to experience anything like that again."

It took Lisa nine years to get on another plane, which she says was one of the hardest things she's ever done. She's since been on eight mission trips that required her to fly and is planning on embarking on more. That being said, she only flies with a buddy who can help her cope with the flashbacks and anxiety that inevitably creep back into her bones during each flight. During a trip to Venezuela, she clamped so tightly onto the leg of the teenager sitting next to her that she left bruises.

While on a trip to Panama, it started storming somewhere near Cuba. Lisa had to close the window shade and ask the nurse-friend sitting next to her if she needed more medication.

"Oh, yes. You do," her friend answered without hesitation.

Lisa's daughter has since become a paramedic who deals with life-and-death situations on a regular basis, but she doesn't talk much about her mother's crash. She has told Lisa that it still disturbs her to think of the plane slamming into a steel pole at ninety-five miles per hour, and she remembers her mother screaming in pain any time she tried to move around the house during her recovery.

Five years after the crash, a memorial was dedicated to honor those who perished in Flight 1420 and recognize the continued courage of the survivors. The monument was built at the Aerospace Education Center across the street from Little Rock National Airport and designed by Larry Thompson, chairman of the art department at Ouachita Baptist University, where nineteen of the passengers were students. During the ribbon-cutting ceremony, some of the survivors sang 'Amazing Grace,' just as they did on that stormy night in 1999.

Some survivors, including Lisa, objected to the pilot's name being included on the memorial. They were outvoted, and because the names were arranged in alphabetical order, the captain is at the top of the list along with the others from Flight 1420 who died.

At that same ceremony, a passenger who was sitting behind her during the accident presented her with a unique memento: a small piece of green plastic from the overhead bin above Lisa's seat that had broken loose during the crash and fallen into the other passenger's purse. The woman knew that Lisa had tried unsuccessfully to convince American Airlines to give her the seat she'd sat in on that ill-fated night. The airline had declined and destroyed the plane after the investigation was complete.

She cried when she saw the bracket. "I have my ticket and other items from Flight 1420 in a cardboard box, but I really wanted something tangible from the plane to touch and feel," she says.

A study done by the Massachusetts Institute of Technology around the same time as the American Airlines crash found some disturbing trends in landings done during the two thousand thunderstorms they profiled at the Dallas-Fort Worth International Airport. In sum, two out of every three pilots were more likely to land during storms when they were running behind schedule, were landing at night, and/or if a plane before them had landed successfully.

In their report, the NTSB determined that the probable causes of this accident were:

> . . . the flight crew's failure to discontinue the approach when severe thunderstorms and their associated hazards to flight operations had moved into the airport area and the crew's failure to ensure that the spoilers had extended after touchdown.

> Contributing to the accident were the flight crew's (1) impaired performance resulting from fatigue and the situational stress associated with the intent to land under the circumstances, (2) continuation of the approach to a landing when the company's maximum crosswind component was exceeded, and (3) use of reverse thrust greater than 1.3 engine pressure ratio after landing.

In addition to the poles being solid rather than being designed to break away, there were a number of airport safety precautions that were not in place that also contributed to the severity of the crash. A thousand-foot buffer had been required at the end of runways since 1988, but the runway used by Flight 1420 that night didn't have one. The runway was built before that date and was grandfathered in under earlier regulations. Additionally, the runway did not yet have an arrestor bed at the end to absorb the weight and slow a plane if it overshot a runway, although one had been approved before the Flight 1420 accident but hadn't yet been installed.

As a result of the Little Rock crash, American Airlines conducted operational and safety reevaluations that included third-party audits of their pilot training and internal organization. Both American Airlines and the Allied Pilots Association have made numerous recommendations including expanding thunderstorm avoidance policies.

WHY WAS I SPARED?

June 7, 2004
Beechcraft Bonanza
Gilcrest, Colorado

Kimberly Clark. Photo credit Ryan Jung.

Summer was just getting underway. Free from the classroom and the grind of studying, Kimberly Clark seized the opportunity to visit her cousins in Colorado. Soon after arriving, her uncle and older cousin brought her to a local airstrip in Longmont where they stored their six-seat, single-engine Beechcraft Bonanza.

The ubiquitous Bonanza is notable in that it has been in continuous production since 1947. It is also the type of plane that crashed on February 3, 1959, killing Buddy Holly,

Ritchie Valens, and J. P. Richardson (The Big Bopper). During the early 1980s, Apple cofounder Steve Wozniak crashed his Beechcraft Bonanza during takeoff and recovered, but guitarist Randy Rhoads, who played with Ozzy Osborne and Quiet Riot died, along with the pilot and another passenger when their plane nicked the band's tour bus before crashing into a tree and nearby home.

On this particular June day, the weather was clear and the temperature was creeping into the low- to mid-nineties. Kimberly's uncle fueled up the plane at the airstrip, and she climbed into the back with her seat facing the tail.

Her uncle was a licensed pilot and her cousin had recently received her student pilot certificate. They cruised over farmland with the Rocky Mountains looming in the distance and circled over their home several times before heading north in the direction of the Greeley-Weld County Airport.

But the serene beauty of the landscape was shattered about twenty minutes into the flight. The engine of the Bonanza suddenly started sputtering and alarms clanged and screamed as the engine lost power.

Kimberly was confused but still felt relatively calm. She listened to her uncle and cousin talking through her aviation headphones. The nearest landing strip was at least ten miles away and too far to be of any use to them during their emergency, so they started looking for the nearest place to land. It was rush hour; using the highway below as a landing strip was out of the question. The only option was the deep-green field to the side of the road.

Her cousin turned around in her seat and told Kimberly to buckle up and hold on. Kimberly didn't fully comprehend what was happening, but she knew it wasn't the right time to ask. It never dawned on her that anything horrible was about to happen. At worst, she figured her aunt would have to pick them up at the field after their impromptu landing.

Her sense of calm didn't last long.

"I can still hear my cousin calling in the mayday and exclaiming that we weren't going to make it to the airstrip. She gave them our coordinates and reported that there were three souls onboard," she says.

As Kimberly peered helplessly through the rear windows, the Bonanza skimmed over power lines. Her uncle wrestled with the powerless plane as it lost momentum and careened downward toward the green sod blanket below. Just as it was about to touch down, the plane rocked from side to side and one of the wings clipped the ground and sent the plane plowing into the earth nose first.

The world went black.

The pain in Kimberly's head was excruciating when she regained consciousness, and she was struck by the seemingly out-of-place smell of freshly cut grass. Unbuckling her seat belt, she could see that the front of the plane was crumbled and crunched inward as she moved to check on the others. She tried to feel her uncle's pulse but there was nothing, and she knew there was no hope from the extensive damage to his head. Then she heard a sickening sound like bubbles being blown in thick chocolate milk coming from the seat where her cousin sat motionless. Despite her labored gurgling, the girl was unconscious but still alive, with a strong pulse.

Kimberly pounded and kicked against the door, frantic to get out of the plane. But the wing was crumpled against the door, preventing it from opening. She let out a blood-curdling scream.

"I've never felt so helpless in my life. I was stuck inside the plane and was unable to help my cousin who desperately needed attention. It was too much to handle," she says.

Drivers who had seen the plane go down stopped and rushed to help. The plane's gas tank had ruptured and the toxic smell of aviation fuel permeated the air with a clear threat of explosion. With all their strength, two men managed to rip the door off and pulled Kimberly out of the heap of crushed metal.

She was carried to a truck and another man sat with her as firefighters and ambulance personnel rushed toward her, but she begged them to help her cousin who was still alive and trapped in the plane.

The doors and the side of the plane were so mangled that firefighters had to cut her cousin out. She was airlifted to the hospital in a helicopter while Kimberly was strapped to a flat board and slipped into a waiting ambulance.

As the ambulance rushed to the hospital, she finally began to process what had just happened. Flooded with overwhelming grief, she cried uncontrollably, the emergency medical technician at her side not knowing how to soothe her. As the doors of the ambulance opened at the emergency room, she could hear the helicopter carrying her cousin landing on the roof.

Kimberly had broken her right elbow and suffered a skull fracture. When she woke in the hospital the morning after the crash, her parents were still driving from Kansas and hadn't yet arrived. She watched the sunrise alone, terribly upset and angry that the world had not stopped. Her uncle had just died in a horrific accident and her cousin was barely hanging on—and yet, the world outside kept moving on as though nothing had happened. It was devastating.

As Kimberly was leaving the hospital the next day, doctors were running a battery of tests on her cousin. She was deemed brain-stem dead, a clinical term for the lack of brain activity.

Her cousin was a registered organ donor and most of her organs and a lot of tissue were harvested. The plane had been moving at about one hundred miles per hour when it slammed to a full stop, and without air bags to lessen the blow, there was severe damage to her cousin's body. Her heart and kidneys were severely bruised and her retinas had detached. Remarkably, she was still able to help others through her death.

When she was finally discharged, Kimberly returned to her uncle's house to pack her belongings. She felt completely vacant inside. Her cousin, to whom she'd been so close, was gone after just twenty-five years on this earth, and would never be back. Her uncle was dead at forty-two years old, leaving his wife and two-year-old twin boys to continue on without him. Kimberly sobbed, furious that she had survived while her uncle and cousin, who had had such full lives with others who depended on them, were no longer around. It wasn't fair.

She tried going to a few therapists, without much luck. The first one began crying after Kimberly recounted her story. Another counselor told her about other people's horror stories, as though it would make her feel better. To placate her parents, she tried yet another therapist, who told her that the worst thing she had ever personally experienced was forgetting to pick up her dry cleaning on time. This therapist, who didn't have a medical license, also suggested that Kimberly start taking drugs for her depression and insomnia. Kimberly and her father took her diagnosis of post-traumatic stress disorder to their family doctor, who was totally opposed to putting her on antidepressants. In the doctor's opinion, it was expected that Kimberly would still be distraught three months after the deaths of her uncle and cousin. In fact, if she were not feeling sad or depressed, there would be more reason to worry. Being told that it was okay to feel this way was actually good for her to hear. The doctor prescribed some mild medication to help her regain a regular sleep pattern.

After some time, Kimberly connected with a therapist who specialized in aviation disasters. Finally, she had found someone who understood where she was coming from and made some progress in healing her emotional wounds.

About a year after her crash, Kimberly started volunteering with the local fire department. Her parents and therapist worried that she might be trying to relive the accident, but giving back to society became the core of her existence and helped Kimberley justify the reasons for her survival. In addition to being a volunteer firefighter, she is also certified as a fire instructor and has invited less fortunate children to her home during Christmas time.

While still in high school and at just eighteen years old, Kimberly took an evening class to become an emergency medical technician and instantly connected with others in the course. She described the volunteers as coming from all walks of life, most having experienced some form of trauma that drove them to become first responders. She might have been the youngest in that program but it allowed her to share her story with an empathic audience. They understood where she was coming from, and she could relate to them too.

Just don't ever say that she was saved for a reason. That expression makes Kimberly angry because it implies that there wasn't a reason for her uncle and cousin to live.

It took seven years for her aunt to broach the subject of the crash with Kimberly. She worried that she might have pulled the plug too early on her daughter, but Kimberly used her emergency training to explain that this was not the case.

"If the ambulance had arrived at the crash site just ten minutes later, my cousin would have died and there would not have been time for anyone to say goodbye," she says.

Four years after the accident, Kimberly and her husband went with her aunt and two sons through the pieces of the plane remaining from the teardown done by investigators. Her aunt stored the dismantled carcass of the plane wreck for years, explaining that she just wasn't ready to get rid of them. Her aunt found rummaging through the wreckage to be cathartic, but it was still not the right timing for Kimberly. Her aunt saved the plane's propeller and the two seats her husband and daughter had last sat in. Kimberly took only her seat belt.

On the tenth anniversary of the accident, for the first time, Kimberly told her family about the anger she carried and explained how it felt to be the sole survivor of a crash that took two lives that were so dear to her. Finally saying those words aloud helped her come to terms with the crash and its aftermath. After a decade of struggling with her emotions, her anger subsided and she feels that some of the burden has lifted from her shoulders.

Today, Kimberly has processed much of the anger and is reenergized through her volunteer efforts. Ten years from the day of that tragedy, she found the exact site where the Bonanza had gone down using the GPS coordinates from the NTSB report. Sitting quietly in the grass with her husband and aunt on that ten-year anniversary was soothing. The stillness was peaceful and she felt like she belonged there, close to her cousin. As she climbed back into the car to leave, the song "Good Riddance" by Green Day was playing on the radio—the same song that had played at her uncle's and cousin's funeral. It was a further sign that she was supposed to be there that day.

Kimberly plucked some grass from the site and gave some to her grandmother. She also shared small tokens, such as the coins she held in her pocket that dreadful day, with other family members. Her grandfather took one of the quarters and made a necklace for her grandmother as a memento of the two that perished.

She doesn't talk about the crash often and many people outside her family and closest friends don't even know that Kimberly was ever in one. When it does come up in conversation, she usually tells a canned version void of emotion. The hardest part for her is when people ask how she felt, such as during the interviews for this book. Her speech becomes thick with raw emotion and it often takes her a few days to pull herself back together.

Kimberly absolutely hates flying and has vowed never to get into a small plane again. When she does fly commercial, she is a nervous wreck and extremely agitated. She usually stocks up on over-the-counter medication that will keep her asleep for as much of the trip as possible. Otherwise, flying is unbearable.

The NTSB report determined that the crash was most likely due to pilot error, but she disagrees. She believes there was a malfunction with the fuel tank that resulted in vapor lock.

Vapor lock occurs when the fuel in the engine changes from liquid to gas. This condition leads to loss of pressure and ultimately engine stall. Fuel can vaporize for a number of reasons, including too much heat from the engine, local climate conditions, a lower boiling point at high altitudes, etc. Restarting an engine that has stalled from vapor lock can be difficult. According to Kimberly, the investigators never found a receipt for the preflight fueling and the gas station at the airstrip lost their database of sales when their computer crashed, leading the investigators to suspect that the plane simply ran out of gas. But she knew her uncle had filled the tanks and remembers the smell of the gas emptying into the sod farm.

On a day when she was feeling particularly low, Kimberly performed an Internet search and found the Plane Crash Survivors group Lindy Philip had launched on Facebook. It was comforting to know that she wasn't alone.

AIRPLANE NUT

November 17, 1990
Weedhopper
Port Neches, Texas

Coda Riley. Photo credit Coda Riley.

When Coda Riley was in high school, he walked to a park and waited for the man who lived across the street to return home. When a car pulled into the driveway, Coda, who was normally quite bashful, scraped together his nerve. He crossed the street and said, "Hi, my name is Coda, and I'm an airplane nut."

Coda recalls, "The man was confused and asked 'What?' So, I said again, 'Hi, my name is Coda and I'm an airplane nut.' He had this funny giggle and right there, Charlie Green Smith took me under his wing."

When his parents divorced, his father moved to a new neighborhood and Coda remembered having seen a small plane in the neighboring driveway a couple of years earlier. He had grown up in southeast Texas surrounded by rice farms, and during the busy season, crop dusters swooped through the skies in all directions. He'd been enthralled with planes for as long as he could remember.

Coda described the sixty-eight-year-old semi-retired optometrist he'd just introduced himself to as being on the shorter side with a potbelly, a crew cut, and a big heart. Doc, as he grew to call him, brought him into his house to meet his wife. That was the start of a close relationship that would impact Coda for the rest of his life.

Doc asked Coda if he would be interested in coming down to the flying field and going for a ride that coming Saturday. He jumped at the opportunity.

They went up in Doc's two-seat Weedhopper, a popular American ultralight that is typically self-assembled from a kit in about twenty-five to thirty hours.

After some instruction, Doc turned the controls over to Coda for an impromptu balloon-popping contest being held at the flying field. Coda was able to steer the ultralight and pop several balloons in a row with the propeller, earning second place and cementing their friendship.

It turned out that Doc had served in World War II as a fighter pilot. According to Coda, he'd completed more than 250 missions in China and Burma, and survived a mid-air collision with another P47 (one of most widely used fighter planes in that war) that knocked one of the wings off his plane, causing the plane to flop into a deadly, downward spiral. When Doc popped the emergency release, the G-force spit him out and deployed the parachute. He was so low that the parachute made one large pendulum swing before Doc tumbled onto the ground, seconds away from becoming a fatality.

Coda had already looked into getting his pilot's license from a traditional flight school but it was too expensive for him and even today he truly believes that he received far better training from the accomplished veteran than he would have anywhere else.

The week following Coda's first flight, Doc called him again and asked if he wanted to come by and help him work on the plane. But when Coda arrived, he didn't load up the truck with tools. Instead, he placed a five-gallon can of fuel into the truck.

"He was trying to see if I'd be willing to put in the work, and apparently I passed the test. The plane didn't really need any work done and we went out flying instead," he says, with an upbeat Texan drawl. "I'd mow his lawn and do whatever I could to help out."

They practiced takeoffs and landings. After a couple of runs, Doc turned to Coda and told him that even though he'd flown with over a thousand students, Coda was the most natural flyer he'd ever met. After just two days of lessons, Doc climbed out of the plane and Coda took his first solo. The teenager wasn't nervous at all because he knew Doc believed in him and wouldn't have let him go up on his own if he didn't have confidence in his abilities.

Doc was also a dealer for the ultralight Weedhopper manufacturer and bought, sold, and traded planes all the time. He taught Coda how to build, assemble, and rebuild ultralights. He also made Coda a dealer too, so that he could start giving his own flying lessons since Doc was getting older and didn't want to bother with teaching anymore.

When Coda turned eighteen, Doc bought him his own ultralight but the factory engine was outdated and weak. When another ultralight was wrecked after someone spun out of control and slammed into a wind sock, they hauled the wreckage to Coda's house and pulled the engine out to use as a replacement for the old one in Coda's plane.

On his first flight out with the replacement engine, Coda decided to visit the neighborhood he'd grown up in, something he'd never done before in a plane. After landing and chatting with the neighbors, he climbed back into his plane and decided to show off a little by flying aggressively, taking sharp turns, banking, and waving the wings at his friends below.

Coda pulled the plane into a steep climb in what he described as "hanging on the prop," but then the unimaginable happened. The engine sputtered and quit. He didn't have enough speed to get the lift he needed to recover. He was on the verge of a stall and plummeting straight down to the ground 150 feet below. He remembers thinking at that point in time, *Oh man, this is going to hurt.*

The next few minutes moved in slow motion. Coda jammed the stick forward in a desperate attempt to force some forward motion, but it was too late. His Weedhopper ran out of altitude and slammed into the ground at a thirty-degree angle.

The plane crumbled and broke into three pieces in a hay field near the house he'd grown up in. The gas tank ruptured and dumped five gallons of fuel on him. He stumbled about fifty feet from the wreck before realizing that he couldn't breathe. At first, he thought he'd had the wind knocked out of him, but it was more than that. He collapsed and lay on his back, gasping for air. The pain was overwhelming.

"My accident had nothing to do with the training I'd gotten or my skill level. It had to do with me hot-dogging and being stupid," says Coda.

A neighbor ran out and took charge. She told Coda to stay still and called an ambulance. When he asked how his plane looked, she told him it was fine. Even in his state of agony, when he rolled over on his side, he was heartbroken to see that it had been severely damaged in the crash.

His mother was working at a nearby convenience store and sped out when she got word about his accident. Coda will never forget watching her fly out of her car while it was still moving. She ran to her son as her car rolled into the ditch on the side of the road.

Neighbors slid Coda onto a sheet of plywood and carried him to the road at the edge of the field where the ambulance was waiting. The technicians strapped him to a backboard and loaded him in.

"The nurses in the emergency room cut off my clothes down to my tighty whities and then had me stand to be X-rayed, rolling my shoulders one way and my hips another. I was in horrible pain and almost passed out. I never thought you'd have to do that kind of thing after a serious back injury," says Coda. "Doc was on vacation in Las Vegas when he got a call about my accident, and he caught the first flight to Texas to check on me."

The hospital finally gave him some Demerol, a welcome relief. The impact of the crash had shattered his L1 vertebrae, reducing its length by three-quarters in an instant and causing it to become a source of pain for years to come.

Coda's father was upset when he got to the hospital. He'd never wanted his son to fly, and, to this day, still doesn't approve. Coda's father had been in the Vietnam War and had been shot at while skimming the tops of trees in a helicopter. The few times that he rode with Coda in a plane, his father clung to the sides of his seat in what Coda described as a death grip.

After the crash, Coda struggled with both physical pain and damage to his pride. A week or so later, he was hanging out with a friend who was dabbling with hypnosis. The amateur hypnotist took him back through every detail and replayed each conversation from the day of the crash in a way that made Coda feel like the traumatic events were happening to someone else. It helped him come to terms with the accident and he began to move on.

Still strapped tightly in a body brace, Coda and Doc spent several weeks rebuilding his plane. Coda wanted to get back to flying, but needed to find someone who would be willing to lend him a plane since his wasn't airworthy. After ruining his Weedhopper, most folks thought that he should take it easy for a while. But it took him only three weeks to get back into the air.

"It wasn't easy sliding into the seat with that brace and I was a little nervous until the wheels left the ground," says Coda. "My parents didn't know what I was doing; my dad would have had a stroke."

For the next six years, Coda flew ultralights. Then he got his general aviation license, followed by instrument training and a commercial pilot's license. He's been flying since 1990 without an accident. Doc, however, wasn't so lucky. While coming in for a landing in an experimental ultralight, the wing came apart, causing the plane to roll and invert. Doc was killed on the spot.

"I didn't even know about it until six days later when I called to check in on him, and the person who answered the phone asked why I hadn't been at the funeral. Everyone thought someone else had called me. I was beside myself and grieved like I'd never grieved before," Coda explains, pain still nibbling at the edge of his words.

While the crash didn't stop Coda from flying, it forever changed how he flies. Today, he makes sure he's aware of the plane's limitations and tries to map out alternatives in case of an emergency.

When Coda needs to stop for gas, for example, he circles the airport first to determine the best place to head to if the engine fails—rivers, a road, or anywhere else that might result in a survivable crash. He says that pilots practice losing an engine all the time, but when a propeller goes *ca-chunk* and freezes in place, the reality of the situation is terrifying.

"The only time I really feel vulnerable is during takeoff when the plane is full of fuel and there's nowhere to go if something goes wrong. If there's a neighborhood or forest at the end of runway, I'm completely at the mercy of the engine until it's climbed a couple thousand feet in the air," says Coda. "There are several airports that I use that I know, if there's an accident, my chance of survival is slim to none. I love what I do, but that's not how I want to go."

It's not just his accident that's given him pause. Over the last twenty-five years, he's lost four or five acquaintances whose planes have burst into flames after crashing. He said that the vast majority of the crashes were due to pilot inexperience, just as his was.

"If I hadn't had my accident, I would probably have died in a plane crash by now. I believe that. That experience made me think more and slowed me down. It was my fault, and if I'd flown that day like I fly now, the crash wouldn't have happened."

Today, he makes his living as a commercial pilot ferrying employees to meetings and other appointments. He also bought his son an ultralight when he was fourteen years old and never hesitates to point out when his son takes unnecessary risks.

Coda is fond of a popular saying in his line of work: *there are old pilots and there are bold pilots, but there aren't too many old, bold pilots.* He considers himself a pilot who was once bold but who today understands the need to respect and finesse aircrafts.

ULTRALIGHTS: FLYING HIGH LIKE A KITE

In the 1970s, as gliding was gaining popularity, a few of the more adventurous souls strapped lawn mower and chain saw motors with propellers to their kites and the concept of the ultralight took off.

Enthusiasts cite not just the affordability of ultralights, but also the fun of flying low with a stick and feeling the wind in their faces. There are also no FAA license or medical certificates required to fly ultralights, although the USUA (the primary ultralight and light sport aviation organization in the United States) highly recommends ten to twenty hours of instruction.

As the number of unregulated ultralights in the air soared (many of them relatively inexpensive do-it-yourself kits assembled in a backyard that don't look much different than the contraptions Orville and Wilbur Wright tried to get airborne at the turn of the twentieth century), so did the number of accidents and fatalities. The FAA responded by passing Federal Aviation Regulations Part 103 (FAR 103) in 1982.

Outside the United States, the term "ultralight" and "microlight" are often used synonymously to also describe single and two-seat recreational planes. However, to qualify as an "ultralight vehicle" under FAR 103, a craft must have only one seat; be used for sport or recreation; weigh 254 pounds or less; carry no more than five gallons of fuel; and travel no faster than fifty-five knots (slightly more than sixty miles per hour). Ultralights can be flown only during daylight hours, and not in clouds or over heavily populated areas or in controlled airspace where there is a greater risk of colliding with another aircraft.

Any ultralight that exceeds these requirements must be registered with the FAA and have a federal airworthiness certificate, and a pilot must have a recreational or private pilot license.

To confuse matters further, people casually refer to ultralights as "planes" even though they are classified as vehicles. If they were actually planes, they would be subject to greater regulation.

Photo credit Associated Press. Photo by J. Walter Green. Firefighters examine the wreckage of an Iberia DC-10 airliner which crash landed at Logan International Airport.

IT HAPPENED SO FAST

December 17, 1973
McDonnell Douglas DC-10
Iberia Airlines
Boston, Massachusetts

Holly Davis. Photo credit unknown.

Susan Tuthill. Photo credit Jay Tuthill.

Holly Davis and Susan Tuthill were passengers on the ill-fated 1973 Iberia Airlines DC-10 flight from Madrid to Boston that crashed the week before Christmas during its landing at Boston's Logan Airport. The DC-10 is a three-engine, wide-belly jet that carries up to 380 people. It was popular during the 1970s with the US Air Force, Federal Express, and commercial airlines for medium- to long-range flights.

Shortly after Susan became a member of the Plane Crash Survivors Facebook group, she saw Holly's post about her crash and knew they were on the same flight. The two became virtual pen pals who emailed back and forth about the crash and their lives afterward, before meeting in-person with their husbands. They continue to stay in touch today.

Susan's husband was also eager to meet Holly, but for different reasons. He wanted to find out more about how she overcame her fear of flying, something his wife had not managed to do.

One day after her twenty-second birthday, Holly was returning to the United States to visit family for the holidays. She was teaching English and studying in Pamplona, Spain, and nearly missed her flight. Because she was running late, her luggage was not placed in the bins with the other bags. Holly hurried onboard and sat in the center aisle, close to the large, cinema-style screen that was used at the time to show in-flight movies.

Susan, too, was living in Spain. She'd spent a semester studying in Seville and her parents had reluctantly let her stay for a year on the condition that she come home at Christmas to visit. Both remember the flight as being uneventful. That is, until the approach to the airport.

It was a stormy day, with snow at the higher altitudes and chilling rain on the ground. Even though it was only three o'clock in the afternoon, Holly had trouble seeing the wing when she looked out the window that was blurred by thick fog and foul weather.

For no discernible reason, Holly was feeling terribly afraid even though she was a seasoned traveler who had already crossed the Atlantic Ocean twelve times. It was unusually dark, and something just felt very wrong. Jittery, she leaned forward in the crash position with her head between her knees.

The woman next to her seemed puzzled and asked, "Are you okay?"

Holly couldn't answer.

Just as the plane approached the runway for landing, it suddenly jarred hard onto the tarmac and bounced into the air. It was thrown back down again, like a rag doll being tossed by a child having a temper tantrum.

The crew on an Air Canada flight taxiing nearby would later tell investigators that they saw the Iberian Airlines flight coming into the runway "desperately low . . . too low to recover." As the aircraft struck piers and the embankment between Boston Harbor and the airport, the main landing gear snapped off and the plane skittered down the runway for about three thousand feet before it came to rest, flames bursting from its side. Inside the cabin, packages and luggage tumbled out of the overhead bins and onto the terrified passengers strapped to their seats.

Susan also remembers an entire overhead bin breaking loose and items flying out of the cabinets in the flight attendant galley in front of her.

Someone opened the emergency door and released the slide. Passengers vacated their seats and started moving toward the exits. Holly recalled that everyone seemed oddly calm after what had just happened, except one of the flight attendants who was nearly hysterical and screeching, "Hurry! Everyone, get out of the plane."

A woman next to Holly started panicking and yelling that there was smoke behind them. Holly grabbed her purse and the box of hand-blown glass Christmas ornaments she'd bought at a flea market in Madrid the day before (eight have survived to date) and hurried toward the escape chute.

Today, Holly often thinks about the young girl, about nine years old, who was sitting a few rows in front of her and was traveling alone. During the harsh landing, the movie screen fell off the wall and hit the child on the head. As passengers evacuated, Holly recalled the girl saying casually, to no one in particular, "That's so weird. I was just in a bus accident. Now, I'm in a plane crash."

She clearly remembers thinking that if the plane was going to explode, running now wouldn't help. Holly was not feeling any sense of doom. At the time, leaving the plane through the emergency exit felt more like an inconvenience—it just seemed like an odd way to get out of the plane.

She struggled onto the inflatable chute. Midway down the slide, a large man lost his balance and toppled onto her. When he pulled her up, she realized they were wading in four inches of Boston Harbor water, just off the runway.

Susan remembers that a fire had erupted just behind her, preventing some passengers from getting out. A nearby emergency door had to be forced open with an ax.

She stumbled out of the plane and into the mud, and watched in horror as the fire spread. Passengers, including a number of children traveling with their families for the holidays, were pulled up through a hole in the roof, and they ran along the wing to escape the burning plane.

The last person to leave the plane was the Spanish pilot. Susan recalls that he was visibly shaken, but was told that because he maintained his composure during the crash, it most likely prevented the accident from being much worse. Thanks at least in part to the pilot's good judgment, there was no loss of life.

It was cold and raining in Boston, not snowing like it was moments before at the higher altitude. The airport emergency crews swung into action, knowing all too well what to do. Only five months earlier, a Delta commuter flight from Vermont came in too low under the cloud cover. Its landing gear hit a seawall, causing a ferocious crash that killed all eighty-nine people onboard and effectively turned the plane into an ash heap.

The reality of what had just happened was still not registering with Holly. She wandered to the nose of the plane where vans waited to carry the bewildered passengers to the airport terminal. It wasn't until that moment that she realized she might actually live through the ordeal. Still, she knew it was a good idea to hustle away from the plane as soon as possible. When she finally stole a glance back, she was stunned to see that the tail of the plane had almost entirely separated from the rest of the fuselage.

Throughout the landing and rescue, Susan's father watched in horror from Logan Airport's observation deck as the plane crashed and burst into flames.

"Now that I'm a parent," she says, "I can't even imagine what he went through. There were no cell phones back then, and he had to wait for two hours to find out if I was still alive."

He also told her that medical personnel had to attend to a number of the family members having anxiety attacks while waiting to hear who had survived the crash.

Eventually, passengers were corralled into a large room—and pandemonium broke loose. A young man asked Holly to hold his bag, but when he never returned, she dropped it somewhere;

today, she wonders if the bag might have contained drugs. Among the survivors was a drunk Spanish teacher, who had both annoyed and entertained fellow passengers during the flight by obliviously singing and blurting out tales of the men she had slept with. Today, Holly is still haunted by the fear she saw on the face of an elderly Spanish woman dressed in black who didn't speak any English and was traveling alone.

An English-speaking doctor was yelling for the injured to move to one side of the room and the uninjured to the other. Holly asked him if he needed a translator because many of the passengers were Spanish and didn't understand what he was saying.

In the confusion, he asked, "Why? Don't you speak English?"

She had, of course, just asked him—in English—if he needed help. The scene in the airport was so chaotic.

It seemed like an eternity before Holly was released to find her parents. The delay, she believes, was due to the airline trying to account for all 153 passengers and fourteen crew members who had survived. Some were in the terminal, but those who were injured had been taken to a local Boston hospital.

Passengers wandered about randomly like dazed zombies. Many were wrapped in blankets with sand and sticks in their hair and dirt on their faces from the plane hitting the ground, gouging the earth, and spattering debris through the holes torn through the plane's skin.

Meanwhile, Susan helped a father clean dirt out of his baby's mouth. A lot of people were injured, which she believes happened during the frantic escape from the burning plane.

While Susan wasn't hurt physically, being so close to death and seeing so many people badly hurt took its toll on her. She says the feeling of having no control while the plane was crashing is something she will never get over.

Photo credit Associated Press. Photo by J. Walter Green. A survivor, covered with mud and foam, walks through the Logan International Airport's rescue center.

Holly's parents, who worked at the Massachusetts Institute of Technology, learned about the crash from a colleague who also told them that the police were asking people to stay away so ambulances could get through. Her mother suggested that they wait until Holly called. Luckily, her dad insisted on going right away. It was highly unlikely that Holly would be able to get her hands on a quarter to call home from a pay booth.

The roads to Logan Airport were deserted. Her mother hopped out of the car at the curb to look for Holly while her dad searched for a nearby place to park. When her mother spotted Holly through the glass plate window, she banged on it to get her attention. Holly remembers looking up and seeing her mother waving with her hand curled like she was grasping something in her palm. She was wearing an old-fashioned rain hat that folded up like an accordion, a sight that added to the surreal disconnection Holly felt with her surroundings. She was roiling with emotion but her parents still had no idea just how serious the crash had been and they were stunned by the number of blanket-swaddled passengers drifting aimlessly in the terminal. When they got home, Holly's father offered her a bourbon, something he'd never done before. She declined, and her father poured himself a tall glass. Holly's sister rushed home from work when she heard the news and was hysterical.

The Information Age was still years away. The local news stations reported no casualties, so her parents had assumed that everything was fine.They were wrong.

When Susan finally found her father, he was in shock and had completely lost his voice. The first thing he could croak was, "We need to call your mother."

Susan's mother had come home from work and was watching *The Merv Griffin Show*, when an emergency announcement came on about the plane crash. Distraught, she called her boss,

Photo credit Associated Press. Photo by Bill Chaplis. The main cabin and wing section of the Iberia DC-10 airliner.

who came over to keep her company until she got news about Susan.

Back at the airport, Susan and her father ended up driving another woman from Maine home after she had called her family to say that she was in the United States to surprise them for Christmas, but had just been in a plane crash. To this day, Susan has never been able to find the women despite several attempts.

When Holly returned to the airport with her brother a few days later, Iberian Airlines made her sign a waiver (which she still regrets doing) before they would release her luggage. Intact suitcases were neatly lined up in a corner, except the nearly unrecognizable, mangled green mess at the end that once was her suitcase. Her bag, with her bikini underwear hanging out from the corner, reeked from a broken bottle of anisette liqueur she'd packed with her clothes.

Holly felt herself starting to unravel at the sight of the suitcase. She was struck by just how close she'd come to dying and kept thinking that it could have been her head that was destroyed instead of the suitcase.

Luggage was an issue for Susan as well. She never got her bags back and they were filled with empty insulin bottles that she needed to refill for her roommate in Spain, who suffered from diabetes. Susan's father had to call everyone with her roommate's last name in the girl's Colorado hometown until they reached her parents and worked out a way for them to get the prescriptions refilled.

Much of Holly's time home during the holidays was spent crying, especially when she thought of getting back on a plane to return to Spain. Her mother called the family doctor for a sedative for the flight back on New Year's Eve, but instead he simply suggested having a few drinks before getting onboard.

When it was time for her to leave, the maimed Iberian Airlines DC-10 was still on the tarmac, shoved to the side by snowplows. Five bourbons later, Holly was crying frantically in the boarding area as jaded businessmen looked on with annoyance. Holly's mother, whom she describes as a woman with a huge personality, turned to the waiting room and said, loud enough to make sure everyone overheard, "See that plane out there? My daughter was on that plane."

Other waiting passengers circled around Holly to comfort her, holding her hand and trying to calm her down. One woman told her that her husband was the pilot and sat with her through the first short leg of the trip to JFK Airport in New York where she was to catch a connecting flight back to Spain. Crazy with fear, she called friends collect from pay phones in New York because she thought she was going to die. She had bought a round-trip ticket with Iberian Airlines and it never dawned on her, or her family, to change the return flight to a different airline.

Through sobs, Holly told the flight attendant what had happened. Within minutes, the pilot came out of the cockpit and sat beside her, telling her that his entire family was onboard this flight and that he was going to give her the best landing of his life.

"I will never know if the pilot was telling me the truth about his family being onboard, but it did help immensely. And yes, the landing was perfect," says Holly.

The plane crash was a true turning point in her life. She was now terrified to fly, but still loved to travel. For the next twenty years, whenever she flew—which was often—she had to drink heavily before she could muster the courage to board a plane. On her honeymoon flight to the Dominican Republic, Holly bawled uncontrollably. Until then, she had wrestled her fear to a tolerable level but was now overcome by how much worse it would be to crash with someone she loved.

Starting a family ramped up her sense of terror even more. A few years after getting married, Holly and her husband were flying with their new baby. She remembers the flight attendant wrenching her daughter from her arms and telling her in a soothing but firm voice that the baby would be happier in Daddy's arms.

"Apparently, I was squeezing the baby so hard that the flight attendant was afraid I was going to hurt her," she recalls.

Her fears invaded other parts of her life as well. Any time Holly went to a carnival or stepped inside an elevator, her imagination took over. She would visualize in excruciating detail, with her heart thumping in her chest, all the things that could go wrong. Once, while watching her daughters on a carnival ride, she had to turn her back so they wouldn't see the tears streaming down her face. Another time during a ski trip, while she rode to the top of the mountain in a gondola, it started rocking back and forth. She clutched to the sides and clearly saw the next day's headline screaming, "Family of three crushed on rocks as gondola falls from broken cable."

Since the crash, Susan developed many of the same fears that plagued Holly. She doesn't remember having these issues before the accident or even after the accident when she boarded her return flight to Spain with Iberian Airlines. This may have been due to a youthful sense of invincibility; however, over time, she became increasingly fearful of flying, elevators, small spaces, and even boats.

For many years, Susan would not let her son fly or participate in any activities she considered dangerous. But she soon realized that it wasn't fair to transfer her fears to him either—so one day, she bought two tickets to Spain.

As the plane was taking off, tears ran down Susan's face as her mortified son stared at her and said, "Mom, you're embarrassing me."

The plane landed without incident.

Because she still lives in Maine, Susan has to revisit the scene of the Iberian crash any time she takes a flight because Logan Airport is the major airport closest to her. She admits the need to push past her fear, "even if it kills me."

"The best part of flying these days is when the plane touches down safely, I am always very thankful," she says.

For sixteen years, Holly continued to drink heavily whenever she flew. She didn't like who she was becoming and finally quit drinking entirely. The first time she flew stone sober, it was incredibly stressful. Rigid and sweating, she hyperventilated and couldn't stop her legs from bouncing uncontrollably. She gripped the armrests like she was going to tear them away from the seats. Takeoffs, landings, and even light turbulence were torture. For another decade, her anxiety levels were out of sight any time she was airborne.

When she turned fifty, she celebrated the occasion by purposefully doing things that were outside her comfort zone, despite her fears of manmade machines that were susceptible to human error. She even enjoyed a helicopter ride that was so smooth it felt like she was sitting on a sofa in her living room.

Finally, after about thirty years of flying in terror, a hypnotherapist cured Holly completely. He had her relive the accident as if it were being shown in a movie theater, making it seem real by describing aspects of the imaginary theater including the popcorn in her lap and the seat she occupied. She then went through the events backward and forward again, revisiting that dreadful December day three times in total.

"The way he did it, I felt safe," she observes. "It was like I was riding a bus. When he was done, it was hard to believe but it worked. Somehow it just did."

Holly is still passionate about travel. To date, she has crossed the Atlantic 116 times and flown over the United States too many times to count. She has visited Hawaii, Mexico, Colombia, Brazil, Japan, China, Australia, and India. But it has been only in the last six years that she's been able to do so without panicking.

Holly doesn't tire of retelling the story of her plane crash. It seems to have defined her as much, if not more, than her other adventures that have included running with the bulls in Spain, driving throughout Morocco, hitchhiking all over Europe, and backpacking around India at the age of sixty.

Inevitably, she is often asked if anyone died in the Iberian crash, and she finds this irritating. "It's as though someone has to die to validate what happened."

Holly is no longer paralyzed by the fear of flying, but there is always an edge of discomfort lurking in the back of her mind. She must assure herself that accidents are possible but one isn't going to happen right now. Other members of her immediate family did not inherit her fears. Ironically, her husband is a glider who is fond of saying that "engines are for sissies" and her brother is a part-time flight instructor. Her nephew was also a commercial pilot at one point in his career.

Surprisingly, Holly has never been afraid of dying. She always figures that she'll go when it is her time. She was raised Catholic, but lapsed. After the accident, she became an agnostic, and after the shooting tragedy at Columbine High School, where her daughter was a sophomore that day, she became a bona fide atheist. Holly no longer believes there is anything that can keep her or her family safe. Planes crash all the time, as do cars. People get killed.

At times, she wonders if her views stem from the accident or if the crash was just one incident in a string of events that made her an atheist. In her heart, she believes her current views are a result of the latter scenario. The crash was just the beginning of her distrust of safety in the world. But that distrust, she believes, has not hampered her in any way . . . with the exception of amusement parks, which she still steers clear of—and that's okay.

WIND SHEAR

Anyone who has ever been on a plane that has experienced wind shear will never forget it. If a plane hits this unstable weather anomaly mid-air, it can feel like it has fallen through a black hole before somehow catching itself and continuing to fly normally. Most dangerous during takeoff and landing, wind shear has been blamed by NASA for crashes that have taken hundreds of lives.

At the time of the Iberian Airlines crash, no instruments existed that had the capability of detecting this phenomenon, which could toss a moving plane in the air with extraordinary violence.

The NTSB report determined that the probable cause of the 1973 Iberian Airlines crash was due to pilot miscalculation, that is, not realizing in poor weather conditions that the plane's rate of descent during landing was accelerating due to wind shear just as he shifted from autopilot to manual controls. As a result of this accident, the NTSB made eight recommendations to the FAA, including the relocation of glide slope transmitters, modifying pilot training, providing informational programs about wind shear, and developing equipment and systems to measure and report the dangerous weather condition.

LIKE FATHER LIKE SON

October 14, 1997
Cessna 150
Albany, NY

Karl Suerig. Photo credit Robin Suerig Holleran.

September 19, 2010
Experimental ultralight
Plattekill, NY

Chris Suerig. Photo credit unknown.

On occasion, there are survivors of plane crashes who walk away completely unscathed, physically and emotionally. Karl and Chris Suerig, father and son (and father and brother of coauthor Robin Suerig Holleran), are both examples of this rare breed. Granted, neither was seriously hurt and there were no fatalities, but given their family history of four plane crashes that resulted in one death and one serious injury, you'd think there might be an inkling of lingering trepidation. But that hasn't been the case.

LIKE FATHER . . .

ORVILLE'S AERIAL TREE-PRUNING BUSINESS

Just two days after singer-songwriter John Denver died while piloting an experimental plane off the coast of California, Karl Suerig came a breath away from losing his life at almost the exact moment his last Suerig grandchild was born—a near-perfect circle of life.

He had joined a friend on a quick trip to Vermont to visit a modular manufacturing company that his friend John intended to hire to build a large home in Greenwich, Connecticut. A few years earlier, Karl had completed a high-end modular townhouse development in nearby Larchmont and was happy to help John finalize his contract negotiations.

The day was overcast but not raining as the two drove to Danbury Airport in northwest Connecticut where John's small Cessna 150 was stored and partially leased to the local flying school, a common practice for offsetting the cost of maintaining a private airplane.

John went into the airport office, while Karl walked around the plane doing a preflight examination, just as he used to years before when he was a Navy flight surgeon. It was some time before John came out the office. Karl assumed that he needed the time to file a flight plan, as is customarily done. But Karl would later learn that this was not the case.

The two men taxied down the runway and took off. The headset system wasn't working, so they had no way of communicating over the loud rumble of the engine other than with hand signals, but that didn't seem like much of an issue for such a short flight.

Soon after getting up in the air, the weather got soupier and soupier. Karl was studying the map on his lap and watching the instrument panel, especially the altimeter that tracked the altitude of the plane. He pointed upward to indicate to his pilot friend that they should try to get above the clouds they were stuck in.

But John shook off the idea, signaling (or so Karl thought) that the cloud cover would only be temporary and the sky would clear in a bit.

Although he hadn't actually asked, it never dawned on Karl that John wasn't licensed to fly with instruments.

"When someone invites you to ride in their car, you don't ask to see a copy of their driver's license or ask if their insurance is up-to-date," Karl says.

About a half hour into the flight, Karl spotted tree tops far closer than they should have been through a brief opening in the cloud cover below. He looked again at the map and was puzzled. If they were flying where he thought they were, there should not have been any hills or trees to worry about.

Again, Karl hooked his finger upward in a clear signal to John that they needed to get higher. John looked at the map and altimeter but continued on the path he'd already set. Karl thought he was being clear but there was no way for him to know for sure if John understood his hand signals when they weren't able to talk through the headset.

Suddenly, there was a loud bang and a wall of tree branches and leaves swallowed the entire windshield just a couple of feet in front of their faces. In the next seconds, the force of the impact with the trees caused the tail to flip up and the plane to slant downward at a sickening angle.

"I know it sounds silly, but all I could think was, *Oh, crap, this is going to hurt,*" Karl says. "I was sure we were going to smack into the ground and die."

By sheer luck, the plane had hit a tree at the top of a cliff. When it came through the other side of the thick foliage, it nosedived down the side of the mountain and somehow John was able to pull the plane back up with the right wing now bent back at an angle not conducive for flying. They were both drenched in sweat. Karl was briefly nauseous and disoriented before pulling himself back to the situation at hand. Because one wing was bent, the plane was crabbing—or flying sideways. John wrestled with the controls to keep it moving forward and prevent it from spinning uncontrollably as the plane was begging to do.

"I don't think we could have made a left turn even if we wanted to," Karl adds.

This wasn't the first time Karl had been in a plane that was crabbing. To earn extra pay while in the Navy, he had to fly a certain number of hours each month. While copiloting a C45 Beechcraft, which Karl described as an old military workhorse of a plane designed in the 1930s, one of the two engines went dead and the plane couldn't maintain altitude. Karl and the other pilot calculated that if they could sneak through a valley between two tall mountains, they would probably be able to make the Durham-Greenville airport. To land the plane that was being pulled to one side by the dead engine (or a damaged wing, as was the situation Karl was currently in), they needed to try the same maneuver used when landing in strong crosswinds. To prevent the plane from drifting to the side during a landing, the nose is pointed into the wind (or toward the side of the still-functioning engine or wing) to counterbalance the drag on the disabled side. As the plane is touching down, the wing on the good side is dipped slightly so that the wheel on that side touches the tarmac first. The friction tugs the plane in that direction and the other wheel is dropped down so that the plane aligns with the center of the runway. "We never called in a mayday back then. We were Navy pilots and thought we were too cool for that," jokes Karl.

John didn't have those ego issues and called in a mayday to the closest airport in Great Barrington, Vermont. The tower asked how extensive the damage was and John reported that the cowling was crushed and they couldn't tell if the plane still had wheels.

The airport didn't have any crash equipment. As long as the plane was still flying and had fuel, they suggested that Karl and John head to Albany International Airport in New York.

Visibility was still poor and none of the instruments on the panel were working, with the exception of a floating compass. With this low-tech device, they knew they were heading west. The tower told them to fly until they saw the Hudson River and then head north, where weather conditions were better.

Albany International cleared the airport and told them to make a standard approach, whereby the plane flies parallel to the runway with the wind, then makes two ninety-degree turns before landing against the wind. As they circled the empty airport, John asked the tower if their plane still had wheels. They did, but the controllers couldn't tell if the wheels were flat or still functioning.

After hitting the tree, John had pushed the throttles to make the engines run hard to get to the airport as soon as possible. Once they got closer and slowed the engines to begin descent, Karl said the plane began to fly better. As they were landing, crash trucks raced next to them on either side of the runway, following the hobbled plane with fire-retardant foam and men ready to jump out and pull Karl and John out of the plane if they flipped.

"When we taxied to an area to park, there was an FAA truck and about thirty people watching us, including television crews. I think they were disappointed when we landed and nothing happened," says Karl. "A friend who lives north of Albany was watching it on television but had no idea it was me."

Karl disembarked and walked around to the front of the plane to look at the cowling and wheels. One of the spectators shouted, "You need to look at the wing."

About three feet from the body of the plane, the wing was crushed flat and the strut was bent. Rather than a flat, smooth surface, the wing's exterior was covered in ripples and irregularities. Spruce and oak branches that had been ripped off when they'd plowed through the trees were sticking up like porcupine quills from the rivets that held the sheets of metal together on the wing.

Karl said the crushed left side of the engine cowling resembled a car bumper after an auto accident. The air intake for the engine was packed solid with leaves. The only reason the engine hadn't failed was that air was being supplied through the small alternative intake created from a two-inch flap on the cowling.

For no explainable reason, the propeller was in pristine condition. Karl still can't figure out how the propeller could have gone untouched after the plane tore through so many trees which caused considerable damage to other parts of the plane.

"People from the FAA were waiting to interview us. I felt pretty stupid. I had experience flying jets in the military and should have known better," Karl says. "In addition to wanting to know what happened, the FAA wanted to make sure I wasn't paying John to fly me somewhere because that would have been an illegal commercial flight and he would have gone to jail."

When the FAA representatives asked John if he'd checked the weather in Danbury, Karl found out that he hadn't. John was actually relying on the weather report he'd seen on television before leaving his home in Connecticut. The investigators shook their heads and laughed, incredulous. The common (and best) practice is to talk with the originating and destination airports to get weather checks for a more complete picture of the conditions for the entire flight being planned.

"I'd assumed a lot. I never thought anyone would do that," Karl adds. "It was a big disruption at Albany Airport. We didn't know it at the time, but they had sent another plane to fly above us and shadow with radar so they could report back to the airport about our progress when we were coming in."

Karl, an emergency room physician in Connecticut, was scheduled to be at work at three o'clock that afternoon. By the time the interviews with the FAA were done, it was noon. He rented a car at the airport and then called his wife.

She listened and said, "I guess it wasn't your time. I'll call the hospital and tell them you're running late."

The emergency room got busy soon after Karl arrived and he didn't think about the accident again, except for when he ran upstairs during a break to meet his son and new granddaughter, who had been born just hours earlier.

About a week later, a handwritten note appeared in his mailbox: *Hey Karl, I've got this great idea. We should go into the aerial tree-pruning business.*

A few days later, Karl put a handwritten note in John's mailbox: *Hey John, Meet me down at the bicycle shop. Orville.*

Rather than give up flying, John earned his commercial license and was an active volunteer with Angel Flight, a nonprofit group that arranges free air transportation for the sick. John made hundreds of flights, shepherding the ill and their families from place to place for medical care.

Then, one afternoon in May 2013, while transporting a brain cancer patient and his wife from the Boston area back to their home in New York, the twin-engine plane John was flying blew up in mid-air, killing all onboard above a sleepy town just west of Albany—not far from where Karl and John had crashed six years earlier.

A little over five years after Karl's accident, his daughter Robin would fall from two thousand feet and crash when the engine of the Cessna she was flying in failed (see "Someone Needs to Buy Me a Lottery Ticket," page 7). Five years later, his son Chris would have a near repeat of circumstances and live to tell the tale.

. . . LIKE SON

TOO MUCH OF AN EXPERIMENTAL FLYING MACHINE

Chris Suerig and his two teenage children met a family friend named Bob at a small airstrip a couple hours north of New

Photo credit Brigette Suerig.

York City on a perfect September afternoon. Bob was a retired entrepreneur who had served in Vietnam and had also been the youngest Brigadier General in the US Air Force after graduating with the first class to go through the newly opened Air Force Academy outside Colorado Springs. His career included a stint as a White House Fellow with General Colin Powell and heading up the Homestead Air Force Base in south Florida before going into business, eventually becoming president of a private security firm.

In his retirement, Bob began flying experimental ultralight planes. Chris believes that after flying military fighter jets, Bob would have been bored with general aviation propeller planes and he found an ultralight powered with something resembling a lawn mower engine more exciting. Bob had two of these planes parked at a small airstrip off the New York State Thruway and offered to take Chris and his kids for rides over the surrounding area.

Ultralights are slow-moving vehicles with one or two seats that are regulated by the FAA but do not require a pilot's license or any special training, although it is highly encouraged, for obvious reasons (See sidebar about ultralights on page 146).

"Ultralights are really hang gliders with a small engine," says Chris. "They're nylon over a metal frame; there's not much to them. The plane we were in actually came with its own parachute. It was that small."

Initially, Chris didn't hesitate about getting into the plane because Bob was such an experienced pilot, but after he saw the flimsy-looking flying machine with the word EXPERIMENTAL stamped in block letters on its side, he became a little apprehensive.

His thirteen-year-old daughter Brigette went up first. Chris watched them take off and the two disappeared from view as they glided over the scenic Minnewaska State Park and buzzed over the lake in front of the Mohonk House for the next forty-five minutes. Derek, Chris's fifteen-year-old son, took the same tour.

Chris was up next.

The sky was a brilliant blue. It was a windless fall day but it was also starting to get late. Chris offered to not go up in the air, even though he loved to fly and wanted to get his own pilot's license at some point in the future.

The two men took off and were about three hundred feet in the air with the whine of the engine over their heads, when Chris heard Bob say over the headset, "We're in trouble."

Chris glanced at Bob with his eyebrows raised, speechless. In that split second, he watched as Bob, still staring intently ahead, shoved the control stick down hard. Because they had not been gaining enough altitude, the ultralight was flying flat and not at an upward angle. It stalled and pancaked on the ground. Instinctually, Chris lifted his feet up off the paper-thin tin floor, just as he heard metal crinkling underfoot.

"It was a very strange sensation. I clearly remember the earth coming up at us; there was no sensation of going down. I'd never seen anything coming at me that fast. It was so unexpected, like being at the top of a ladder that collapses. There was no time to think," Chris says. "We hit so hard, the right wheel completely snapped off."

Lopsided with a wheel missing and the engine still revving, the ultralight spun sharply to the left and slammed into a hangar at the end of the runway.

Bob shut off the engine and said, as if he forgot that Chris was still sitting next to him, "I just totaled my plane. That doesn't make sense."

Bob continued to ignore everything around him and spoke out loud to himself as he tried to diagnose what had happened. Chris had to ask him twice if he was okay, before he got a curt, "I'm fine."

Totally rattled, Chris fumbled with his cell phone but couldn't figure out how to call his kids who he was certain had just watched the plane go down. He scrambled to the top of the hill and saw Derek running toward him.

Derek had been filming Chris's ride with a hand-held video camera, but was so flustered when he watched the plane drop from the sky that he accidentally deleted the recording. Brigette was reading a book in their truck when Derek yelled that the plane had just crashed as he took off running.

"Very funny," she said, not even looking up.

As nonchalantly as he could, Chris walked over to his kids and said, probably not all that convincingly, "We're okay, we just re-landed."

Until that day, Bob said he had never lost control of a plane in his life. He thought the crash might have been caused by humidity or some other environmental condition. Chris, on the other hand, wonders if it was a weight issue. At over six feet tall, Chris weighed about two hundred pounds. Bob, also over six feet, was probably in the 240-pound range.

"The plane just didn't have enough power to get off the ground with both of us in it. I guess, there's a chance my knee might have pushed against the throttle, but I don't think so," he says.

They grabbed a dolly from Chris's truck, shoved it under where the wheel had broken off, and dragged the plane to the hangar by tying a rope to the back of the truck. The four of them went out to lunch at a local diner and then went on their own ways.

"My body was sore as hell, like I'd been in a car accident," Chris says. "But it was certainly better than the alternative. It takes considerable force for a wheel to snap off like that. I have no idea how I didn't hurt my spine or do more damage to my body."

However, Chris didn't want his kids to worry, so he put on a calm face. As soon as he was alone, he called his sister Robin.

With the thrill of adrenaline still zinging in his voice, he said, "Now, we're three for three." Meaning that now his father, Karl, his sister, Robin, and he had all survived different plane crashes—quite the family legacy.

Chris doesn't think about the incident too often, except every time he drives up the New York State Thruway on the way to a family cabin (which is frequently in the warmer months) and sees the airstrip off to the side of the road.

And, to this day, his kids never found it necessary to tell their mother what had happened that September afternoon. She will continue to be blissfully unaware, that is, until she reads this book.

AFTERMATH

The repercussions of a plane crash dramatically affect those who survive it and reverberate deeply among family members, close friends, first responders, and others who touch the disaster in some way and are jolted out the normalcy of everyday routine.

On July 17, 1996, Heidi Snow got a phone call from her mother asking, "Was Michel on the TWA flight to Paris?"

Michel Breistroff had just left. TWA Flight 800 exploded off the coast of Long Island about fifteen minutes after taking off from JFK Airport in New York.

Heidi dialed TWA repeatedly but couldn't get through. She hoped he was on another fight but deep down she knew the truth. She'd seen her fiancé's printed ticket before he'd left.

A friend went to the airport to get more information while Heidi stayed close to the phone, but there was total pandemonium at JFK Airport. The airlines had no protocol at that time for handling a situation like this. Her friend got the news that Michel was on the roster before Heidi or Michel's family did.

A nearby Ramada Inn became a makeshift headquarters for those waiting to hear news about what had happened to the 230 souls aboard the flight to Europe. At first, airline officials wouldn't let Heidi in because she wasn't yet married to Michel and was not next of kin. But she finally found a sympathetic employee and was given an entrance badge to join the others.

"I was the last person to see Michel alive. I had to call his family in Paris to tell them that their son was onboard the plane that had just gone down. It was one of the hardest things I've ever had to do," says Heidi.

His parents flew to New York but after a few days it became obvious that nothing could be done so they returned to France to grieve.

Heidi's life became surreal. She realized that the world around her was still moving forward. She was expected to go home too, and continue her life as if nothing had happened. An impossibility. She was twenty-four years old and her twenty-something friends couldn't relate to what she was going through as much as they tried.

"I would start to cry and felt disconnected from the people around me. It was very lonely," she says. "My perspective on life changed the day of the crash and would never be the same. I needed to talk to someone who understood."

At the Ramada Inn, Heidi had met New York City Mayor Rudolph Giuliani. She later called him to find a plane crash support group but there wasn't one.

He directed her to a meeting with family members of Pan Am Flight 103 that was destroyed by a terrorist bomb on December 21, 1988. All 259 people onboard were killed, along with eleven people on the ground in Lockerbie, Scotland, who were hit by sections of the plane that rained down upon them.

There, she met a woman who had lost her fiancé onboard Pan Am Flight 103, and Heidi instantly had someone to talk to and cry with, someone who truly understood what she was going through.

At the end of the two-hour meeting, Heidi passed around a pad of paper and asked the Pan Am group members to write down their names, contact information, and a brief description of their circumstances so that she could match them up with others who were grieving after the recent TWA disaster whom she knew were facing similar challenges.

This networking model used for matching TWA families with those from the Pan Am group worked so well that it continues to be the foundation of the organization ACCESS (AirCraft Casualty Emotional Support Services) that Heidi formed after losing Michel.

The day after the first ACCESS board meeting, Swiss Air 111 flying from New York to Geneva crashed into the Atlantic Ocean. At the time, there seemed to be a major airline accident about once every two years.

"No one should have to deal with that type of loss alone," Heidi says.

ACCESS and the hundreds of compassionate volunteer grief mentors who have survived or lost loved ones in private, military, and commercial airplane crashes and other aviation tragedies provide non-denominational, non-political, long-term support to others struck with similar losses.

In Heidi's experience, those who have lost loved ones in private crashes often seek services for longer periods of time. Such services are especially needed on the anniversary of private plane crashes because there are no organized family support groups or memorial services for small, non-commercial tragedies. ACCESS has also formed strong partnerships with other organizations and provides information about additional available resources.

"Matching is critical in helping people cope with feelings of hopelessness, isolation, despair, and confusion that often follows an air disaster," adds Heidi. "We match mothers to mothers, siblings to siblings, spouses to spouses and other relationships lost. Our goal is to minimize the risk of post-traumatic stress syndrome, long-term depression, and social difficulties that can haunt people for months or years afterward. We've fielded thousands of calls from people who've lost loved ones as long ago as 1958 who still need someone to talk to."

In an effort to continually improve their services, ACCESS follows up with those who have sought their help. The organization also has consulted with a number of forward-thinking airlines to help train their first responders.

Heidi points out that suffering loss as a result of an airplane crash differs from other tragedies in that the remaining friends and family have a hard time making sense of how the accident could have happened. Facts are frequently slow to be released, if ever, questions are frequently left unanswered. The information that does get passed on can be difficult to absorb, a process that is further muddled if conspiracy theories or terrorist acts are involved. Family and close friends often fret about their loved one's last moments, wondering if they knew they were going to die and if they were scared.

"It's not uncommon to hear those left behind say they wish they could turn back the clock so they could have stopped them from getting on the plane," adds Heidi. "Michel was in New York visiting me and wasn't feeling well the morning he left. In hindsight, I could have tried to convince him to stay because he was sick, but he had a ticket and went to JFK to catch his flight."

Fortunately, others like Heidi have managed to turn their grief into positive change.

Karen Eckert was waiting for her sister, Beverly Eckert, who was flying from Newark, New Jersey, to Buffalo, New York, on February 12, 2009. A news flash rolled across the television announcing that a small plane had just gone down, but when it changed to a commercial flight from Newark, she screamed.

The police had no information but suggested that Karen could go to the airport. Not waiting for confirmation, she called her three other siblings. This was the second time the family would experience the trauma of an aviation disaster.

Beverly was traveling to Buffalo for a family event to celebrate what would have been her husband's fifty-eighth birthday and was in the process of organizing her high school reunion. She was also scheduled to present a scholarship in her husband's honor to Canisius High School. Shawn, whom she had been with since she was sixteen years old, had died eight years earlier when an airline hijacked by Al Qaeda terrorists rammed into the South Tower of the World Trade Center. Unable to escape from the upper floors, Shawn stayed on the phone with Beverly for the last thirty minutes of his life before the building collapsed. None of his remains were ever recovered.

Beverly became one of the prominent voices for the families of 9/11 victims. Her anger was initially focused on the terrorists' ability to board a commercial flight with box cutters. She became involved in many aspects of the investigation that followed, with a focus on intelligence reform to help prevent future terrorist attacks. Beverly left her professional career in the insurance industry after her husband died. She preferred to live a more purposeful life, spending years in Washington, DC, working to make air travel safer. She also volunteered with Habitat for Humanity (another way to honor her late husband who enjoyed rehabilitating older homes) and at a local soup kitchen, helped children learn to read, and provided companionship to area senior citizens. The week before she died, Beverly had been to the White House to meet President Obama.

During the briefings that followed the Buffalo crash, Karen started to realize that something wasn't right. Initially, she assumed that the accident was weather-related but she sensed there was anger with the NTSB head investigator who addressed the group. In the audience, there was a former Continental Airlines pilot and father of one of the victims who challenged some of the details being reported. Karen took copious notes, not knowing that they would become important later on.

Karen says that during the formal NTSB hearings in May, three months after the disaster, all those in the room, including members of Congress, were alarmed when they heard crash details, including the inexperience of the cockpit crew. The pilot of the flight, which was billed as a Continental Airlines connection rather than a flight by subcontractor Colgan Airlines, slowed the plane to a dangerously low speed as it neared the airport and then, in response to the plane's warnings, made a series of mistakes that resulted in an aerodynamic stall.

The aircraft, shown in an NTSB simulation, swung back and forth at terrifying angles before flipping, slamming into the roof of a home in a quiet residential neighborhood, and bursting into a fiery ball. Forty-nine people onboard and a man who lived in the destroyed house were killed. Miraculously, the man's wife and daughter were in a different room at the time of the crash and managed to escape.

According to the official report, the pilots were exhausted. The first officer was making only $16,000 per year, a base salary that would later be dubbed during the hearings as "food-stamp wages." The copilot, unable to live on her own near her base in Newark, had flown overnight on a cargo plane from the West Coast where she lived with her mother. Before the crash, she was recorded as saying that she felt ill and shouldn't be flying but needed the overtime pay.

The audience also learned that, before deciding to become a pilot, the captain had been a stock boy and attended a "pay-to-work" flight school operated by the now-defunct Gulfstream International Airlines.

To apply for a commercial airline position in 2009, a pilot needed only 250 hours of airtime and only had to be eighteen years of age. Granted, the airline would provide additional training for a particular airplane, but the baseline requirements for many regional carriers was shockingly low. In a sense, employment with a regional was comparable to a low-paying internship that might evolve into a better-paying position. But, in the interim, passenger lives were at stake.

It was noted that Captain Marvin Renslow had failed five check rides (comparable to a driver's test for pilots). During the ensuing litigation, Colgan emails (that had not been provided during the NTSB investigation) disclosed that the airline initially denied him the promotion needed to fly the type of plane he commanded that fateful February night. This caused many to question whether regional airlines that compete for lucrative contracts with larger carriers were cutting corners and putting profits ahead of safety.

After listening to all of the deficiencies, one parent in the audience accused Colgan of murdering their child.

Scott Mauer lost his daughter in the Buffalo crash too. She was on her way to upstate New York to meet her boyfriend and attend his brother's wedding. She'd just told her parents at Christmas that he was the man of her dreams.

"At first, the Families of Continental Flight 3407 group just wanted to understand what happened. The FBI and CIA were involved in the beginning to make sure that the crash wasn't a terrorist act, which was very confusing. When we peeled back the onion, it was obvious just how wrong things were. We ended up coming together to make a difference," says Scott.

He explained that many regional airlines operated at minimal standards, most of which were written in the 1950s and not relevant to the equipment in use today. Major airlines realized in the 1970s after a series of horrific, fatal accidents that they needed to improve and go beyond basic requirements, but many of the regional airlines did not follow suit.

"It was the worst day of my life when that plane went down. I miss my daughter terribly," Scott says.

Karen and her sister Susan Bourque retired from their careers and became core members of the Families of Continental Flight 3407 group with Scott and several others. They didn't want to go to Washington because they knew how hard it had been for Beverly during her 9/11 efforts. But they also knew that Beverly would have wanted them to correct a wrong.

"Beverly cherished the people who were important to her, and she made them feel like they were a gift in her life," says Karen. "She opened the path for us to deal with her death."

As it turned out, the FAA Reauthorization Bill was up for renewal and many of the safety regulations had not been updated in fifty years. According to Karen, the Buffalo crash was the sixth fatal regional airline crash in the same number of years. The timing was right for change.

"I used to worry about security and terrorism but after learning about what was happening on regional flights, I started to worry about pilot training and the amount of sleep the flight crew was getting," adds Karen. "I always have an overwhelming urge to knock on the cockpit door and ask questions. The public shouldn't have to worry that way."

Families of Continental Flight 3407 fought hard to improve the minimum qualifications and training for commercial pilots, establish standards to minimize fatigue, and create a database for airline employers to check pilot backgrounds.

At last, smaller regional airlines are required to provide CRM (see page 104) to their flight crews, and tickets must disclose the names of the regional carrier that is flying on behalf of larger airlines, so that passengers know exactly who is behind the controls.

"We are supporters of commercial aviation. That's why we are fighting so hard to ensure there is one level of safety whether it's a major airline or a regional airline," says Karen.

Scott also said he got a call after recent elections that there are plans being made to roll back parts of the new FAA guidelines they'd fought so hard to improve.

"Airlines should be able to make money and fly safely," he says.

While safety regulations pose their own set of challenges, the legal process following an aviation disaster can be daunting too. Litigation will never bring back the lives lost and asking survivors and victims' families to relive the events of a crash and its aftermath is traumatizing.

Doug Latto, who has been practicing aviation law for twenty-three years, explained that financial recoveries for families can also differ considerably in the United States depending on the state, and that there is a movement to have aviation litigation moved to federal court as a means of limiting damages.

He said clients are often surprised to find out that when the NTSB convenes an investigation, only the airline (if it was a commercial accident), plane and component manufacturers, and any others that have relevant technical expertise are allowed to be a part of the investigative process. The argument is that these parties have the most knowledge about the aircraft and any related components. While this might be true, the existing process still prevents family members from hiring their own experts to participate. Only after the NTSB releases factual reports can the family hire an expert to break down the engine as part of any ongoing litigation.

The vast majority of accidents are attributed to pilot error, and Doug points out that manufacturers participating in an investigation aren't exactly a disinterested third party. To the contrary, it would be in their best interests not to identify a mechanical problem as a potential cause of an accident.

In addition, there are protections afforded to the aviation industry that may seem Kafka-esque to those who have never been through a similar legal process. For one, the General Aviation Revitalization Act passed in 1994 shields general aviation aircraft manufacturers from expensive and protracted product liability claims that were being blamed for a steep decline in the industry and a corresponding loss of jobs. The statute protects manufacturers of small planes from any liability on a product that is eighteen years or older, even if negligence is involved, based on the notion that a defect would have been evident before that time.

This is significant because about three-quarters of the general-aviation aircraft in the United States are older than this. In fact, the average single-engine plane registered with the FAA is forty-one years old.

A *USA Today* four-part investigative report disclosed some additional disturbing facts. Crash-resistant fuel tanks have been available since the early 1970s. When the Army installed them in military planes, it dramatically reduced their fatality rate. But general aviation manufacturers have successfully argued that installing crash-resistant fuel tanks in their planes, at several thousand dollars per tank, is cost prohibitive.

The newspaper also printed that some aircraft issues that persist today have been known for decades. For example, since the 1960s, it has been common fact that bolts holding pilot seats in Cessna planes can fail, causing the seat to slide back so far that pilots can't reach the controls. The military was advised to replace the part in 1975 but civilian owners weren't notified until 1983. And when the FAA contemplated mandatory replacement in 1987, Cessna said it didn't have the capability to produce enough replacement parts. Instead, the company offered an alternative solution that ended up being so difficult to install that most Cessna owners never bothered. After one incident severely burned three victims, a Florida jury awarded them $480,000,000. The court called Cessna's lackadaisical response to the seat-sliding problem as "neither timely nor appropriate."

One engine component manufacturer knew that a defective part in a carburetor was an issue since the 1970s, but told the FAA that the engine problems were due to pilots using automotive gas instead of aviation fuel. This deception, which one Alaska judge called "outrageous," resulted in a number of multimillion-dollar settlements.

Tens of thousands of private planes and helicopters are exempt from current safety standards such as shoulder belts and bird-resistant windshields because they are grandfathered in under the standards in place when the plane was originally designed—not produced. Even if design modifications have been made over time, the exemption stands. For example, the popular Piper Cherokees and Cessna Skyhawks are required to only meet safety standards from the 1960s and 1970s because that's when the planes were originally designed.

All of this lends credence to the ongoing debate of whether there exists a conflict of interest when the NTSB relies on general aviation manufacturers to help determine the cause of a crash. It may also explain why important facts that sometimes contradict NTSB conclusions aren't unearthed until later litigation.

Each year, the NTSB investigates approximately 1,500 general aviation accidents that cause about 475 deaths of pilots and passengers. These incidents are frequently attributed to lack of "adequate knowledge, skills, or recurrent training to fly safely particularly in questionable weather conditions."

Regardless of a pilot's proficiency level and training, there will always be an element of pilot error in any plane crash. That said, it does not excuse the underlying cause of a crisis. A pilot might not respond appropriately to a mechanical failure but it may have been a defective part that created the situation in the first place.

While commercial flight is considered a safe mode of transportation, general aviation is not. A recent NTSB report tagged the general aviation accident rate as six times higher than small commuter flights and forty times higher than airlines. Many life insurance companies armed with mountains of supporting statistics exclude death in a private plane for general aviation pilots as a payable event, along with motorcycle accidents, acts of war, racecar driving, and other risky activities.

Aviation safety relies on continuous improvement. Unfortunately, it sometimes takes a disaster, or near disaster, to prompt change, which is why flying today is still safer than it has ever been in the past.

SAFETY MANAGEMENT SYSTEMS

As a direct result of the Buffalo crash, the US federal government passed rules in January 2015 requiring passenger and cargo airlines to have safety management systems (SMS) in place by 2018. These systems collect and analyze safety data to spot troubling trends, proactively mitigate risks, and prevent accidents. According to the FAA, the causes of 123 accidents between 2001 and 2010 could have been identified beforehand if the planes were equipped with SMS.

The concept of SMS has been around for some time. In 2006, the UN agency International Civil Aviation Organization recommended that countries throughout the world develop SMS rules for their local airlines. Most commercial airlines in developed countries already have some form of SMS in place to identify, control, and continually assess potential risks, but the new rules formalize requirements and expectations of data-driven safety programs and strengthen airline safety culture.

TIPS FOR SURVIVING A PLANE CRASH

Many plane crashes are survivable. Because each crash is different, there is no magic formula as to where to sit, but there are a few basics that you can follow to improve your chances of surviving should you ever find yourself in the position of being in a plane that's going down unexpectedly:

- Keep your shoes and socks on in case there's a fire and/or you have to run across glass or sharp objects.
- Wear lace-up comfortable shoes that you can move fast in. High heels can puncture an inflatable slide and sandals can fall off, leaving your feet exposed to injury.
- Wear long pants and long sleeves made of natural fibers like cotton and wool, not highly flammable synthetics that can melt into your skin if there's a fire.
- Practice unbuckling your seat belt so it becomes part of your muscle memory. When real emergencies happen and people panic, they sometimes look for a release button, like you would on an automobile seat belt, and waste valuable time fiddling with the seat belt latch.
- Know exactly where the closest (and the second and third closest) exit is. Plane crashes are often complicated by darkness, bad weather, and smoke—and it's not uncommon for an exit to be blocked or damaged after a crash. If you

know that the exit is exactly five rows ahead, you can feel your way up to it even when it's difficult to see.

- Know how the emergency doors open.
- Listen to flight attendants while they are giving safety instructions and study the safety card in the pocket in front of your seat.
- Get into the brace position during an emergency landing. Put the seat in an upright position and, if there is room, grab your ankles or lower legs and place your face down (not to the side) in your lap. If there isn't enough room between seats to do this, put your forehead on the seat in front or you with your dominant hand on the back of your head first and cover it with your non-dominant hand. It's important to protect the stronger hand that will be needed to release seat belts and possibly operate other emergency equipment.
- Keep your knees and feet together, with your feet flat on the floor. Your feet should be slightly behind your knees. If possible, store your luggage under the seat in front of you. All of this will help reduce the risk of breaking the bones in your legs and feet, which are common during plane crashes and can prevent you from making a speedy exit.
- If the flight crew is giving instructions for evacuating, then certainly follow them. But if they're unable to do so, don't wait. They may be incapacitated or in shock. Get out of the plane as quickly and calmly as possible. Panic and hysteria will not help. If you can, try to help any other passengers who may be frozen in place.
- Do not try to use pillows or blankets as extra cushions when a plane is going down. They can get in the way when you're trying to get out.
- Leave your belongings behind. Nothing is worth the risk that the extra time collecting your stuff will lower your chance of survival (and those trying to get out behind you). It might

sound like common sense, but you'd be surprised at how many people actually do stop during an emergency to pick up their bags and gifts they're bringing home.

- Be hyperaware during takeoffs and landings. This is when the vast majority of accidents occur.
- Try to sit in an aisle seat within five rows of an exit.
- Once on the ground, get away from the plane. They are filled with aviation fuel and the chances of fire and explosion are real. However, do not leave the scene entirely. Rescue crews will be sent and they will need to do a head count to make sure everyone is accounted for. If you are hurt or fall or pass out, you also want to be near enough that someone can easily find you.
- For those who fly on small, private planes, the July/August 2013 issue of the *FAA Safety Briefing* included an article "What Would MacGyver Do? A Look into Aviation Survival" that provides tips that might have been seen on the popular 1980s television series about a secret agent that used everyday materials to solve complex problems. In addition, the FAA Civil Aerospace Medical Institute offers postcrash survival courses that are available to the public.

ON DEPARTURE

We are truly grateful to those who opened their hearts and took the time to share their traumatic stories. Surviving a plane crash and its aftermath is a unique part of each person's personal journey; their thoughtful accounts are courageous and inspiring.

Life is a gift. When you've come close to losing it—whether in an accident, due to an illness, or in some other circumstance—you learn firsthand just how precious each day is.

We are grateful to Skyhorse Publishing for believing in our project and giving us first-time authors an opportunity to share our experiences and those of others.

MEET THE AUTHORS

ROBIN SUERIG HOLLERAN

Robin is a full-time freelance writer whose work has appeared in local, regional, and national consumer publications, business journals, and on internal websites and other communications vehicles used by her business clients. An earlier account of her plane crash won Honorable Mention at the 75th *Writer's Digest* Writing Competition.

It took nearly two years for Robin to physically recover from a broken back and other injuries sustained during her plane crash but she has since returned to a lifelong passion for traveling off the beaten path. Nearly five years after that life-altering day, she reached the top of Mt. Kilimanjaro with a women's group from the Red Ribbon Foundation to raise money for AIDS research in honor of her late stepbrother.

Robin is a single mother with three children, a dog, two cats, a lizard, and a lot of energy.

LINDY PHILIP

Writing a book has always been on Lindy's bucket list. The opportunity came about when she met Robin through the Plane Crash Survivors Facebook group.

Lindy is an artist and avid gardener. She lives on Vancouver Island, Canada, on an acreage with her husband (another plane crash survivor), two active teenagers, a dog, and two goats. Her plane crash experience has not stopped Lindy from flying and traveling, although she does prefer the smoother ride of larger airplanes.

REFERENCES

"46 Die as a New York-Bound Jetliner Crashes in Spain." *The New York Times*, September 14, 1982. www.nytimes.com/1982/09/14/world/no-headline-149829.html.

AirDisaster. http://www.airdisaster.com/.

American Airlines Flight 1420 Family. www.jrily.com/1420Family/index.htm.

"Bad Place to Bring Plane Down." *Enterprise-Journal*, October 21, 1977.

Burkhart, Teri. "Con escaped in confusion." *Tri-City Herald*, December 29, 1978. http://mattcampsblog.blogspot.com/2007/08/1978-portland-plane-crash.html.

"Busy Night at Hospital." *Enterprise-Journal*, October 21, 1997.

Civil Aviation Accident Investigation Commission. *Technical Report: Accident occurred on September 13th, 1982, to McDonell Douglas DC-10-30-CF aircraft, reg. n. EC-DEG, at Malaga Airport.* http://fomento.es/NR/rdonlyres/24451F6D-3AC8-40F0-8D74-F611D46EF39C/14154/1982_031_A_Spantax_English1.pdf.

"Crash hero prefers freedom to prison." *Desert News*, December 30, 1978. http://news.google.com/newspapers?id=xeNSAAAAIBAJ&sjid=aYADAAAAIBAJ&pg=5265,5106241&hl=en.

"Control on pilots urged." *New Zealand Herald*, 1977.

"The Day That Changed My Life! The Crash of Spantax Flight 995." *Stand Clear and Watch the Closing Doors!*, August 24, 2013. https://tampagr8guy.wordpress.com/2013/08/24/the-day-that-changed-my-life-the-crash-of-spantax-flight-995/.

Diehl, Alan E., PhD. *Does Cockpit Management Training Reduce Aircrew Errors?* Canberra, Australia: 22nd Annual Seminar of the International Society of Air Safety Investigators, November 1991. www.crm-devel.org/resources/paper/diehl.htm.

"Eight attempts to land fail, says son." *New Zealand Herald*, 1977.

Engle, Michael. "Culture in the Cockpit—CRM in a Multicultural World." *Journal of Air Transportation World Wide* (2000). www.skybrary.aero/bookshelf/books/2503.pdf

"EU lifts flight ban on 4 Indonesian cargo planes." *The Jakarta Post*, April 23, 2011. www.thejakartapost.com/news/2011/04/23/eu-lifts-flight-ban-4-indonesian-cargo-airplanes.html.

"Europe: A Grisly Triptych of Disasters." *Time*, September 27, 1982. http://content.time.com/time/magazine/article/0,9171,925742,00.html.

"FAA to require airlines to use data to prevent accidents." *The New York Times*, January 7, 2015. www.nytimes.com/aponline/2015/01/07/us/politics/ap-us-airlines-safety-data.html.

Federal Aviation Administration. *FAA Administrator's Fact Book.* June 2012. www.faa.gov/about/office_org/headquarters_offices/aba/admin_factbook/media/201206.pdf.

Federal Aviation Administration. *Press Release—FAA Final Rule Requires Safety Management System for Airlines.* January 7, 2015. https://www.faa.gov/news/press_releases/news_story.cfm?newsId=18094.

"Flight Conditions Beyond License." *New Zealand Herald*, 1977.

"Focused On Failure (Fatal Fixation)." *Mayday*. National Geographic Television, March 11, 2003.

Frank, Thomas. "Investigations: post-crash files in small planes cost 600 lives." *USA Today*, October 27, 2014.

Frank, Thomas. "Safety last: Lies and cover-ups mask roots of small-plane carnage." *USA Today*, June 12, 2014.

Frank, Thomas. "Small and dangerous: FAA rules mean new planes with old technology." *USA Today*, June 14, 2014.

Frank, Thomas. "Unchecked carnage: NTSB probes are skimpy for small-aircraft crashes." *USA Today*, June 12, 2014.

Harris, Richard. "Beware the 'Safe' Airplane." *In Flight USA*, July 2007. http://home.iwichita.com/rh1/eddy/Safe_Airplane_NOT.htm.

Harter, Andrea. "Flight 1420 plaintiff sobbingly testifies about her distress." *Arkansas Democrat-Gazette*, April 11, 2001. www.arkansasonline.com/news/2001/apr/11/flight-1420-plaintiff-sobbingly-testifies-about-he/.

Harter, Andrea. "'Forever linked' through Flight 1420: Crash survivors deadicate memorial to dead, honor heroes of night." *Arkansas*

Democrat-Gazette, June 2, 2004. www.arkansasonline.com/news/2004/jun/02/forever-linked-through-flight-1420/.

"Helped by radio before crash." *New Zealand Herald*, 1977.

"His Thank You Was Goodby." *Milwaukee Sentinel*, December 30, 1978. https://news.google.com/newspapers?id=w4RQAAAAIBAJ&sjid=-RIEAAAAIBAJ&pg=6699,7012817&hl=en.

"Inquest Told of Bid to Talk Aircraft Down." *New Zealand Herald*, January 24, 1979, pg. 14.

Karl, Dick. "Gear Up: Back in Life's Saddle." *Flying*, June 20, 2011. www.flyingmag.com/pilots-places/pilots-adventures-more/back-life's-saddle.

KingRey, Ed. "Close-Up: United Airlines Flight 173—A Controller's Account." *AVweb*, January 17, 1999. www.avweb.com/news/safety/183017-1.html.

Lallanilla, Marc. "How to Survive a Plane Crash." *LiveScience*, June 8, 2013. www.livescience.com/38015-how-to-survive-a-plane-crash-ntsb.html.

Mathewson, Nicole. "Experienced pilot in plane crash." *The Press*. February 28, 2014. http://www.stuff.co.nz/the-press/news/north-canterbury/9773599/Experienced-pilot-in-plane-crash.

Mollard, Angela. "Motorists drag pair from fire." *Auckland Star*.

Morse, Caroline. "How to Survive a Plane Crash." *The Huffington Post*. February 18, 2014. www.huffingtonpost.com/smartertravel/how-to-survive-a-plane-cr_b_4810499.html.

National Aeronautics and Space Administration. *Making the Skies Safe from Windshear.* April 22, 2008. http://www.nasa.gov/centers/langley/news/factsheets/Windshear.html.

National Transportation Safety Board. *Improve General Aviation Safety.* May 2012. http://ntsb.gov/safety/mwl/Pages/mwl5_2012.aspx#reports.

National Transportation Safety Board. *NTSB 2015 Most Wanted Trasnportation Safety Improvements.* 2015. http://www.ntsb.gov/safety/mwl/Documents/MWL_2015_Factsheet_01.pdf.

National Transportation Safety Board. *NTSB/A-83-45 Safety Recommendations.* July 12, 1983. http://www.ntsb.gov/safety/safety-recs/recletters/A83_45.pdf.

National Transportation Safety Board. *NTSB/AAR-01-02 Aircraft Accident Report.* October 23, 2001. http://www.ntsb.gov/investigations/AccidentReports/Reports/AAR0102.pdf.

National Transportation Safety Board. *NTSB/AAR-74-14 Aircraft Accident Report.* November 8, 1974. http://www.ntsb.gov/investigations/AccidentReports/Reports/AAR7414.pdf.

National Transportation Safety Board. *NTSB/AAR-76-17 Aircraft Accident Report.* August 18, 1976. http://www.ntsb.gov/investigations/AccidentReports/Reports/AAR7617.pdf.

National Transportation Safety Board. *NTSB/AAR-78-06 Aircraft Accident Report.* June 19, 1978.

National Transportation Safety Board. *NTSB/AAR-79-07 Aircraft Accident Report.* June 7, 1979. http://www.ntsb.gov/investigations/AccidentReports/Reports/AAR7907.pdf.

National Transportation Safety Board. *NTSB/AAR-78-06 Aircraft Accident Report.* June 19, 1978.

National Transportation Safety Board. *NTSB/AAR-85-06 Aircraft Accident Report.* July 10, 1985. http://www.ntsb.gov/investigations/AccidentReports/Pages/AAR8506.aspx.

National Transportation Safety Board. *NTSB/ATL06LA028 Probable Cause Report.* December 27, 2005. http://www.ntsb.gov/about/employment/_layouts/ntsb.aviation/brief2.aspx?ev_id=20060105X00013&ntsbno=ATL06LA028&akey=1.

National Transportation Safety Board. *NTSB/ATL84FA080 Probable Cause Report.* January 11, 1984. http://www.ntsb.gov/about/employment/_layouts/ntsb.aviation/brief.aspx?ev_id=20001214X38504&key=1.

National Transportation Safety Board. *NTSB/DEN04FA087 Probable Cause Report.* June 7, 2004. http://www.ntsb.gov/_layouts/ntsb.aviation/brief.aspx?ev_id=20040615X00793&key=1.

Nesbit, Sharon. "One terrifying night." *Portland Tribune*, December 29, 2008. http://portlandtribune.com/component/content/article?id=80695.

"Piper Lance." *Professional Pilots Rumor Network.* http://www.pprune.org/private-flying/396500-piper-lance.html.

"Piper Lance Turbo T Tail." *Piper Forum.* http://www.piperforum.com/showthread.php?t=670.

"Prayers mark 10th anniversary of Ouachita plane-crash tragedy." *Baptist News Global*, June 2, 2009. http://baptistnewsglobal.com/archives/item/4120-prayers-mark-10th-anniversary-of-ouachita-plane-crash-tragedy.

"Prisoner aids jet passengers, escapes." *Wilmington Morning Star*, December 30, 1978. http://news.google.com/newspapers?id=M7ssAAAAIBAJ&sjid=NBMEAAAAIBAJ&pg=4625,6610474&hl=en.

Rhoda, D. A. and M. L. Pawlak. *An Assessment of Thunderstorm Penetrations and Deviations by Commercial Aircraft in the Terminal Area.* Lexington, MA: Massachusetts Institute of Technology, June, 3, 1999. https://www.ll.mit.edu/mission/aviation/publications/publication-files/nasa-reports/Rhoda_1999_NASA-A2_WW-10087.pdf.

Scanlon, Matthew. "Build Easy to Assemble Low-Cost Ultralight Aircraft From Kits." *Mother Earth News*, February/March 1997. http://www.motherearthnews.com/diy/low-cost-ultralight-aircraft-zmaz97fmzgoe.aspx.

Siebert, Al, PhD. "Survivors of the United Flight 173 Crash." *THRIVEnet*, January 1999. http://www.thrivenet.com/stories/stories99/stry9901.shtml.

Safer Healthcare. "Effecting positive behavioral and cultural change . . . Crew Resource Management." http://www.saferhealthcare.com/crew-resource-management/crew-resource-management-healthcare/.

United States Ultralight Association. http://www.usua.org.

Wald, Matthew L. "3 Safety Features Absent At Airport in Little Rock." *The New York Times*, June 9, 1999. http://www.nytimes.com/1999/06/09/us/3-safety-features-absent-at-airport-in-little-rock.html.

Whitaker, Sterling. "37 Years Ago: Lynyrd Skynyrd's Plane Crashes, Killing Three Band Members." *Ultimate Classic Rock*, October 20, 2014. http://ultimateclassicrock.com/lynyrd-skynyrd-plane-crashes.

Williamson, Mike. "Investigation into airplane crash continues." *Enterprise-Journal*, October 24, 1977.

Williamson, Mike. "Plane crash questions lingering unanswered." *Enterprise-Journal*, November 3, 1977.

Williamson, Mike. "Skynyrd crewman nearly nixed plane." *Enterprise-Journal*, October 25, 1977.

Whipple, Julie. "The 1978 Burnside Airplane Crash." *Portland Monthly*, December 1, 2013. http://www.portlandmonthlymag.com/articles/2013/12/1/the-1978-burnside-airplane-crash-december-2013.

Woods, Sabrina. "What Would MacGyver Do? A Look into Aviation Survival Equipment." *FAA Safety Briefing*, July/August 2013. http://www.faa.gov/news/safety_briefing/2013/media/JulAug2013.pdf

Yan, Holly and Karla Cripps. "Missing plane and air disasters: How bad was 2014?" *CNN*. January 29, 2014. http://www.cnn.com/2014/12/29/travel/aviation-year-in-review.